T0259696

FAUNA ENTOMOLOGICA SCANDINAVICA
Volume 2 1974

The Sesiidae (Lepidoptera) of Fennoscandia and Denmark

by

M. Fibiger & N. P. Kristensen

Colour-plates by Roland Johansson

SCANDINAVIAN SCIENCE PRESS LTD.
Gadstrup . Denmark

Copyright for the World
Scandinavian Science Press Ltd.

Edited by
Societas Entomologica Scandinavica

Editor of this volume
Leif Lyneborg

World list abbreviation
Fauna ent. scand.

Printed by
Vinderup Bogtrykkeri A/S,
7830 Vinderup, Denmark

ISBN 87-87491-02-8

Contents

Introduction

The present work deals with the 17 species of the lepidopterous family Sesiidae occurring in Fennoscandia and Denmark. Also treated are a few additional species known from the adjacent N.W. European faunas and probably await discovery in the area under treatment. In accordance with editorial policy, British species are given full treatment, whether or not their occurrence in Fennoscandia and Denmark is likely.

All species are keyed, diagnosed and illustrated in colour. The genitalia of both sexes are described and figured. Scandinavian Sesiidae are generally easily identifiable on superficial characters, but the information on genital structure may be of value in identifying aberrant or damaged specimens. Information on distinguishing characters of larvae and pupae are provided for many species.

Brief description of local and total distribution of the species is given. The catalogue provides a more detailed survey of the known local distribution, based on the standard zootopographical division of the area.

Information on bionomics and collecting techniques is given in greater detail than will probably be the case with most other volumes in this series. Adult Sesiids are rarely encountered in nature, and the much needed further work on Sesiid faunistics is consequently dependent on knowledge of larval habits and rearing techniques.

The sections on the bionomics of each species, collecting and rearing techniques as well as the catalogue were prepared by M. Fibiger. The remaining parts were written by N. P. Kristensen, who also undertook the final editing of the entire manuscript in collaboration with L. Lyneborg, managing editor of the series.

We are indebted to numerous colleagues for information and loan of material: Dr. W. Dierl, Zool. Staatssammlung, Munich; Lic. P. Douwes, Zool. Inst., Lund; Mr. D. S. Fletcher, British Museum (Nat. Hist.), London; Dr. H. J. Hannemann, Zool. Museum, Humboldt-Universität, Berlin; Mr. J. Jacobsen, Copenhagen; Mr. S. Kaaber, Århus; Dr. J. Kaisila, Zool. Museum, Helsinki; Dr. F. Kasy, Naturhistorisches Museum, Vienna; Mr. S. Kerppola, Helsinki; Dr. H. Krogerus, Helsinki; Dr. C. Naumann, Inst. für angewandte Zoologie, Universität Bonn; Mr. I. Norgaard, Copenhagen; Mr. M. Opheim, Zool. Museum, Oslo; Mr. E. Schmidt-Nielsen, Naturhistorisk Museum, Århus; Mr. P. Skou, Troense; Mr. I. Svensson, Österslöv; Dr. E. Urbahn, Zehdenick; Dr. N. L. Wolff, Zool. Museum, Copenhagen and Mr. M. Vuola, Helsinki. We thank Mr. K. L. Elsman and Mrs. G. Lyneborg for preparing the line drawings for publication, Mr. G. Brovad for photographic work and the council of the Amateur Entomologists' Society for permission to reproduce illustrations published in their leaflet no. 18.

A very special thanks is due to Mr. R. Johansson, Växjö, who gave us much advice, lent us many important specimens and, above all, prepared the coloured plates with the highest degree of artistic ability and scientific accuracy.

State of knowledge of the Sesiidae of Fennoscandia and Denmark

Although the "tinaeoid" nature of Sesiidae has been recognized for more than a half century, the family is still being considered "Macrolepidoptera" by students of this group. Collection of Sesiids has consequently been carried out by a great number of amateur lepidopterists in all Scandinavian countries. However, since it can be pursued efficiently only by a knowledgeable search for immature stages, members of the family remain generally relatively rare in private as well as in public collections, and the faunistic records of this group are far more inadequate than is the case with most other "Macrolepidoptera". The results of some recent collecting by specialists suggest that several species now considered local or rare will eventually prove to be much more widespread or common.

Sesiidae have been dealt with in all handbooks on Scandinavian "Macrolepidoptera", the most recent treatments being Gullander (1963), Hoffmeyer (1960), Nordström & Wahlgren (1941) and Valle (1937). The distribution of the species in Fennoscandia has been comprehensively described and largely mapped by Nordström et al. (1961). On the present treatment the distributional data in the last-mentioned work and in Hoffmeyer (1960) have been supplemented with a number of published and unpublished records, the Danish data being largely derived from unpublished maps prepared by S. Kaaber.

Systematic position and diagnostic characters in Sesiidae

The family Sesiidae belongs to the large suborder Ditrysia (characterized by females having separate copulatory and ovipository orifices) but little precise evidence bearing on its immediate relationships is available. In all modern classifications Sesiids are assigned to a position somewhere in the tinaeoid complex of family groups. This complex may be somewhat differently conceived, but in any case it is merely an assemblage of overall primitive Ditrysia. Within the tinaeoid complex the Sesiidae belong to a grade in which pupal abdominal spines, adult ocelli and spinosity of humeral plate are retained, and the proboscis has not yet acquired broad semicircular sclerites externally (Börner, 1939, 1959). They conform with some other families of this grade, viz. the Glyphipterigidae, Douglasiidae and Heliodinidae, in the prominence of the ocelli, great reduction of maxillary palpi and trend towards a simple female frenulum (Common, 1969). It is possible that a true phylogenetic relationship with these families exists. Common, followed by Bradley et al. (1972), assigns them all to a superfamily Yponomeutoidea, but relationships with Epermeniidae and Yponomeutidae also contained therein are not obvious. Brock (1971), after separating the Glyphipterigid subfamilies Glyphipteriginae and Choreutinae at family level, erected a superfamily for the Sesiidae + Choreutidae (and tentatively also Hyblaeidae as well as the little-known Dudgeonidae), but precise evidence in favour of this arrangement is not given.

The Sesiidae are medium-sized to rather small (but never minute) moths with rather narrow wings. All Scandinavian species are immediately recognizable by having at least the hindwings transparent (except for the margins and the veins); usually the forewing too is largely transparent, and because of the usually prominent pattern of light (red, yellow or white) bands on the dark abdomen a striking resemblance to hymenopterans is produced. The few other N. European moths with extensive wing-transparencies (Hemaris (Sphingidae), Thyris (Thyrididae)) have entirely different facies.

Structurally the family is characterized primarily by a unique wing-coupling apparatus in addition to the usual frenulum/ retinaculum (see Common, 1969; Naumann, 1971): The forewing hind margin is folded down, the hindwing anterior margin is folded up, and both folds are equipped with a series of recurved spines which interlock (Fig. 1). In the hindwing, according to the morphogenetic studies by Comstock (in Beutenmüller, 1901) the medial stem is displaced to the anterior margin of the discal cell (i.e., it is basally inseparable from Sc+R_1) and the vein running to the apex is Rs+M_1, not Rs alone. Recourse to these characters may be necessary for identifying some extra-European Sesiidae,

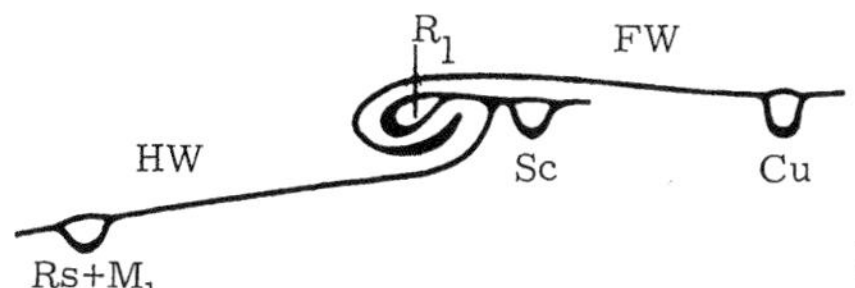

Fig. 1. Cross-section of wing-coupling apparatus (redrawn after Engelhardt).

which either have entirely scale-covered wings or occur together with other wasplike moths (notably Arctiidae-Syntominae).

Some other specialized features of adult Sesiids were stated by Naumann (1971: 44) to be confined to this family: Loss of forewing medial stem; single female frenulum bristle; reduction of maxillary palps; position and great development of ocelli. However, these characters are all encountered in several other lepidopterous groups and cannot define the Sesiidae, but as mentioned above, some of them may assist in determining the systematic position of the family.

According to MacKay (1968) Sesiid larvae may be identified by the following characters: Borers; body pale and without pattern except on pronotum; thoracic segments often enlarged; four upper stemmata arranged in a trapezoid and remote from lower two; the two halves of the cranium usually meeting at a deep and acute angle posteriorly; bristle L1 on meso- and metathorax on its own elevation ("pinaculum") about equidistant from bristles SD1 and L2, with L3 closer than SD1 or L2 to L1 and posterior and usually somewhat dorsal to it; L1 and L2 always adjacent on abdominal segments I-VIII (Fig. 2); prolegs with crotchets uniserial and usually arranged in two transverse bands.

Naumann (l. c.) states that the pupae of Sesiidae are characterized by the abdominal spine-rows, but as mentioned earlier this is a feature of general occurrence in pupae of lower Lepidoptera. In practice, Sesiid pupae may usually be identified as such by the approximately circular arrangement of stout spines on the cremaster as well as by the situation in which they occur.

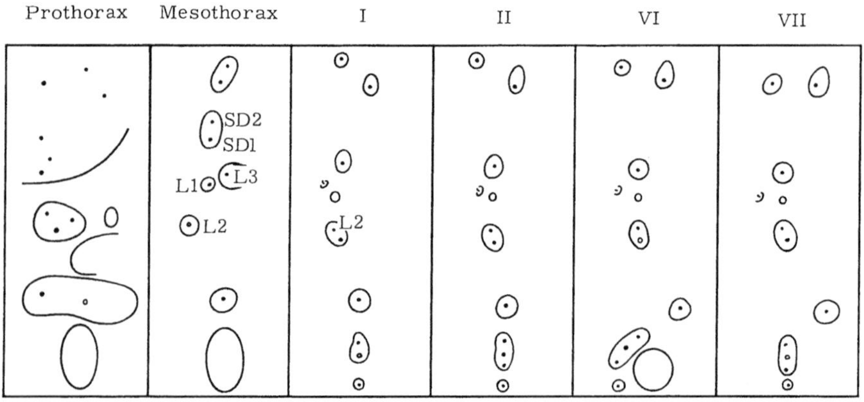

Fig. 2. Bristle "map" of a generalized Sesiid larva (redrawn after McKay).

Taxonomic characters in Sesiidae

ADULTS

HEAD. Characters of the head-capsule, visual organs and mouthparts have so far played little role in Sesiid taxonomy. The proboscis has become desclerotized and nonfunctional independently in several lineages within the family. The texture and coloration of the cephalic scale-covering is of some significance; for descriptive purposes the following terms may be applied to the more or less distinct scale-groups: *facial* (anteromedially), *laterofacial* (along the inner margins of the compound eyes), *vertical* (dorsally), *postvertical* and *postocular* (on the posterior surface of the head-capsule, medially and laterally respectively). The scale-covering of the two basal labial palp segments may be adpressed or tuft-like raised; that of the terminal segment is generally almost smooth.

The antennae, as usual in Lepidoptera, show some sexual dimorphism, males always having an equipment of setae projecting from the scale-covering. Primitively within the family the antennae are *filiform* (distally tapering), but in the majority (subfamily Sesiinae) they are *clavate* (subapically thickened). Both types may show surface-increasing specializations: they may be *bipectinate* (with double branches on each segment) or *pectinate/dentate* (with a single, more or less prominent, projection on each segment). The clavate antennae of Sesiinae have a most characteristic apical hair-pencil (Fig. 3).

THORAX. (Fig. 4). Colour patterns of the thoracic scale-covering are often of taxonomic importance. As in most ditrysian Lepidoptera the short pronotum carries a pair of dorsal, stalked protuberances, the *patagia*, the scales or hairs of which form the *collar*. The forewing base carries a large, platelike, stalked appendage, the *tegula*, which largely covers the lateral part of the mesonotum and the scale-covering of which is often characteristically coloured, particularly along the margins. The coloration of the scale- and hair-covering of the legs is sometimes distinctive, particularly that of the hind tibia.

WINGS. (Fig. 4). The taxonomically most important variation in the forewing (abbreviated FW) venation concerns the interrelationships of R1 and R2, which

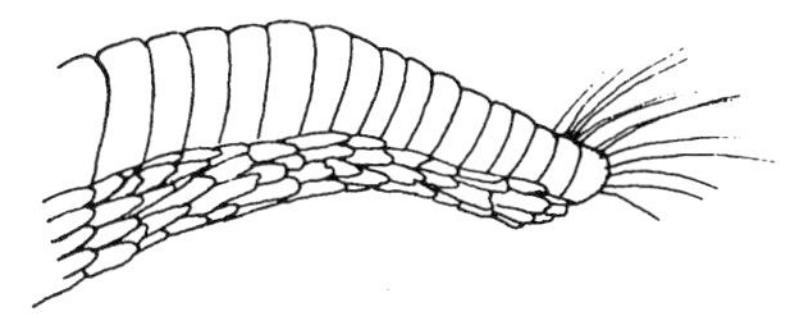

Fig. 3. Apex of antenna of female Sesia.

may be parallel, converging or confluent, as well as the position of the radial fork (stalked R4 and R5) and the number (2 or 3) of medial branches present. The hindwing (HW) first postcellular vein, A1, is reduced in the subfamily Sesiinae. The reduction is not immediately obvious, however, since its position is indicated by a scale-covered line.

A particularly noteworthy feature of the Sesiid wings is, of course, the transparency, which may be more or less extensive, but which in the European species is always present, at least on the HW. The wing margins and the veins are generally covered with pigmented scales. A prominent scale-covered spot, the *discal spot*, is located at the outer end of the discal cell in the FW; in the HW it is less developed or entirely absent. In the FW the two transparent areas separated by the scale-covered cubital stem have sometimes been termed the "longitudinal field" and "cuneiform field" (German: "Längsfeld" and "Keilfeld", respectively), but here the self-explanatory terms *anterior transparent area* (ATA) and *posterior transparent area* (PTA) will be used. The transparent area distal to the discal spot is the *external transparent area* (ETA) whereas the outermost field, covered with pigmented scales, is the *apical area*. The scale-covering of the transparent areas is much less dense than in the unmodified areas, and is modified in one or two different ways (Kristensen, in preparation): either the scales are pigmented, but largely deciduous, coming loose during early flight (Sesiinae-Sesiini) or they are colourless, but persist during adult life.

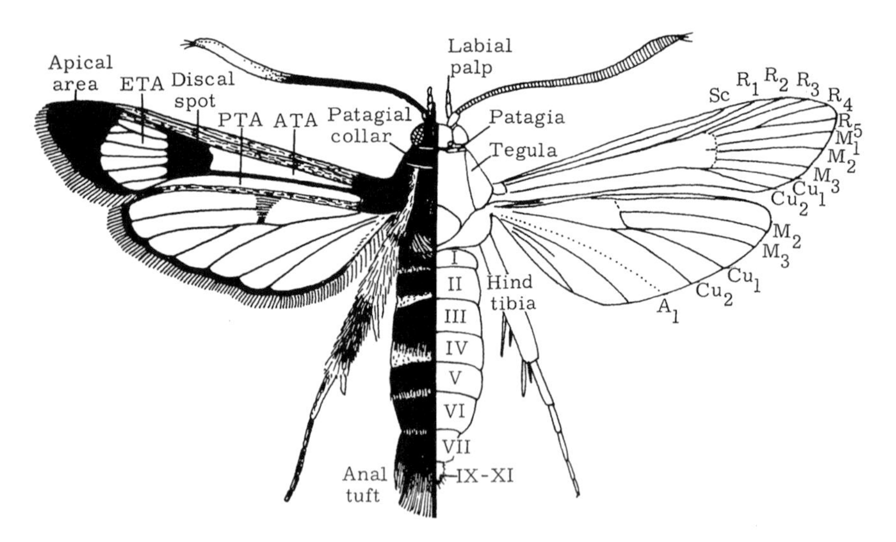

Fig. 4. Female of Aegeria scoliaeformis (Bkh.), showing terminology of main external characters.

ABDOMEN. (Fig. 4). The scale-covering of the abdomen generally has a most characteristic pattern of yellow, white or red markings. In most cases only the dorsal pattern will be dealt with in the descriptions; this pattern comprises transverse bands and sometimes also longitudinal rows of spots. In some species the number of transverse bands permits an easy distinction between the sexes, males having one more band than females; this is related to the fact that in the females the postabdominal modifications affect one segment more than in males. It should be pointed out, that the number of abdominal bands is not always absolutely fixed within each species (even disregarding artifacts, see below); aberrant specimens may have bands in addition to the normal ones or lack one or more of the latters. Coloration is also occasionally subject to variation; thus some of the "red-banded" Aegeria-species may exceptionally occur in yellow-banded forms.

The tergites and sternites of segment VII (in females) and VIII (in males) have a more or less prominent covering of long hair-scales, and a pair of large tufts are inserted on narrow sclerotizations in the pleural membranes. The entire "brush" of long hair-scales, comprising the tergo/ sternal as well as the lateral groups, is here termed the *anal tuft*.

GENITALIA. For a full survey of the nomenclature of the genitalia of both sexes reference should be made to Tuxen (1970). Naumann's (1971) selection of terms are used in the present work.

The male genitalia of primitive Sesiidae largely conform to a pattern widespread among ditrysian Lepidoptera (Fig. 5). The *uncus* is distinctly demarcated from the *tegumen* and the latter from the *vinculum*. In higher Sesiinae (Aegeriini and most Paranthrenini) the uncus and tegumen become fused. Most Aegeriini (Fig. 6) have a conspicuous *scopula androconalis*. This scopula is an elongated, membranous sac situated at the apex of the tegumino-uncal complex and proximally continuous with the socii. Along its margins it is densely covered with specialized sensillae. In the Paranthrenini the uncus is much elongated (and the teguminal part of the complex reduced) and laterally densely covered with setae; this probably represents a functional analogy to the scopula androconalis. The anal cone often has a distinct ventral sclerotization, the *subscaphium*. A *gnathos* is absent in the Tinthiinae, but distinct, with paired processes, in all Sesiinae. In Sesiinae-Aegeriini the gnathos processes are long flat flaps; in the midline between these flaps is a variably shaped duplicature, the *crista gnathi*, which may appear as an anterior continuation of the median part of the subscaphium. The *valvae* are elongate, rounded; in primitive groups either with scattered ordinary setae (Tinthiinae) or with clusters of setae apically and on a central elevation (some Sesiinae-Sesiini). In higher Sesiinae the valvae have a strikingly specialized sensory equipment. Proximally in the ventral part there is an elevation, the *crista sacculi*, covered with strong sensillae which may be pointed stout setae, flattened scales or intermediate types. The crista may have a "diagonal" course, i.e., it may be proximally directed towards the middle of the valve hinge. In the Paranthrenini, apart from the crista, the valvae have a peripheral covering of long setae, the dorso-proximal of which are apically scale-like expanded and scalloped; similarly specialized dorso-proximal setae have evolved independently in some Sesiini. In the Aegeriini the valvae have a *sensory field* extensively covered with specialized sensillae arranged in rows. The alveoles of the sensillae in each row are connected by cuticular ridges (visible only at high magnification). The individual sensillum is apically somewhat widened, bent and distinctly bifid (Naumann, 1971: "Sinneshaare vom Synanthedon-Typ"); sensillae of the same type occur on the scopula androconalis. All Sesiinae have a pair of posteriorly directed *processi vinculi* on the vinculum, near the lower articulation of the valve. The *saccus* is moderately developed in primitive forms, but may be much elongated. The diaphragm contains a sclerotization, the *anellus*, which forms a ventral support for the phallus. In some Tinthiinae the anellus is almost closed dorsally, but usually its extension is restricted to the ventral part of the diaphragm, and it could therefore be termed a *fultura inferior* or a *juxta*. Its postero-lateral corners are often distinctly elongated posteriorly and in higher Sesiinae they are bent towards the phallus. The invaginated part of the diaphragm (i.e., the wall facing the phallus) is the *manica*. The *phallus (aedeagus*, *penis*, etc.) of Tinthiinae and Sesiinae-Sesiini has an anterior coecum, since the ductus ejaculatorius enters the sclerotized tube some distance behind its anterior end; in higher Sesiinae it enters at the anterior apex. The apical portion of the phallus may bear characteristic keels, spines, etc., whereas the equipment of internal cornuti rarely is of obvious taxonomic significance.

In the female (Fig. 7), segment VIII is sclerotized on the lateral surfaces whereas the venter of the segment is membranous to a variable extent. The copulatory orifice, *ostium bursae*, is located mid-ventrally, either immediately behind sternum VII or somewhat more posteriorly. A sclerotization in venter VIII in front of or behind the ostium bursae is termed *lamella antevaginalis* and *lamella postvaginalis* respectively. The *ductus bursae* may be sclerotized to a variable extent. The portion of the ductus immediately inside the ostium is often distinctively shaped or sclerotized and is then termed the *antrum* (Naumann uses this term to designate the entire part of the ductus between the ostium and the origin of the ductus seminalis). The *corpus bursae* usually is a simple membranous sac; only exceptionally is it furnished with characteristic ribs or sclerotizations, *signa*.

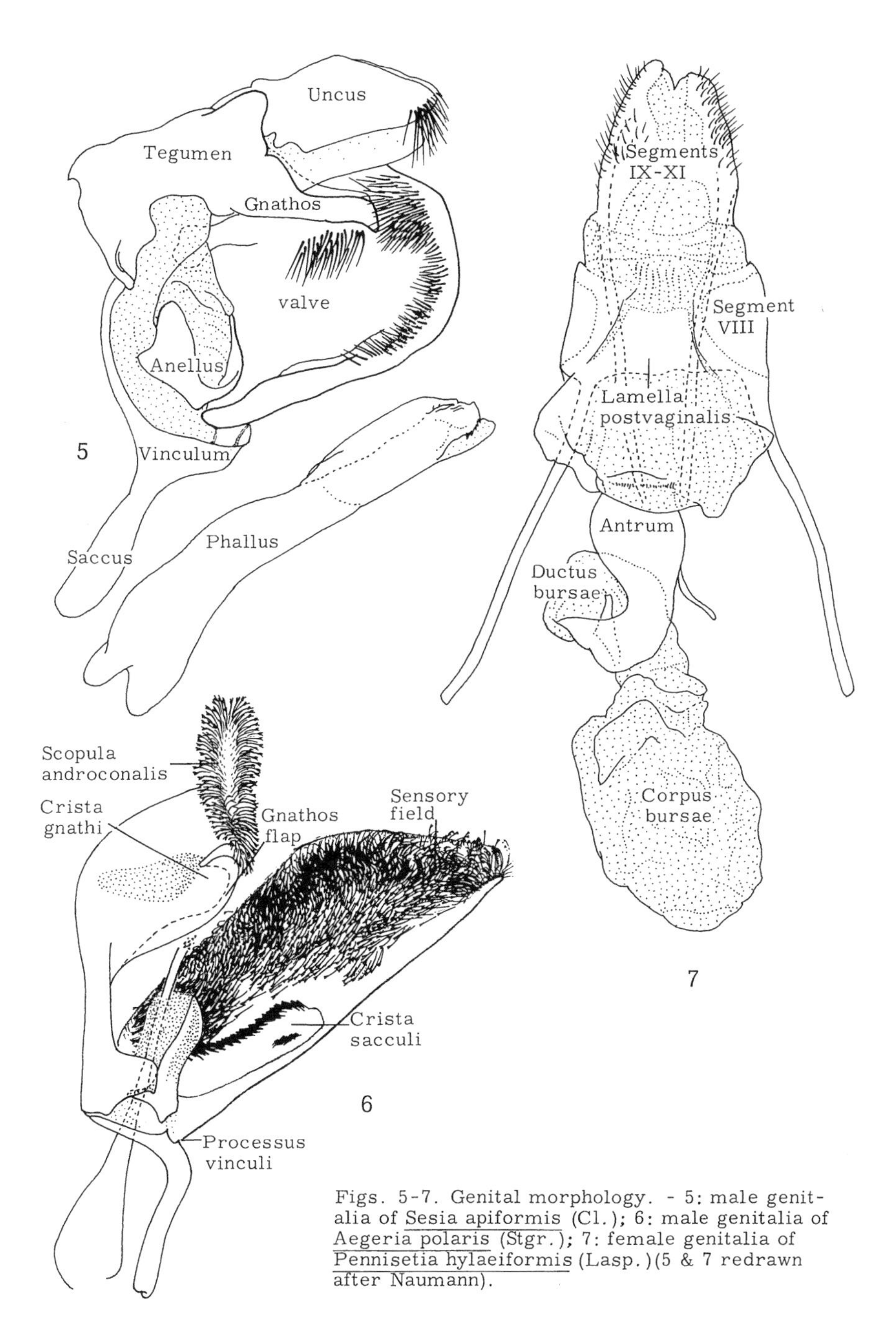

Figs. 5-7. Genital morphology. - 5: male genitalia of Sesia apiformis (Cl.); 6: male genitalia of Aegeria polaris (Stgr.); 7: female genitalia of Pennisetia hylaeiformis (Lasp.)(5 & 7 redrawn after Naumann).

TECHNICAL NOTES. When examining the coloration of preserved specimens the possibility of artifacts should be taken into consideration. Dried Sesiids very easily become greasy, whereby lightly coloured scales are darkened. The scales of the thorax and face are easily rubbed off, thereby giving a unicolorous appearance to naturally contrasting areas. Some species have very faintly indicated abdominal colour patterns, which may be partly obliterated by loss of scales during flight.

Upon dissection the male genitalia of Sesiidae are very easily arranged in the conventional manner, i.e., with the valvae spread out. However, as pointed out by Naumann (1971) this method does not yield reproducible results, since the uncus-tegumen complex may be distorted in different ways. Naumann recommends carefully cutting off the valvae and embedding these along with the genital segment proper, which is placed on the side; the phallus may be removed or not. Alternatively the genital segment may be divided by a horizontal cut between vinculum and tegumen; the dorsal part is then mounted on the side and the ventral part with the valvae spread. Most male genitalia preparations for the figures in this book were made according to the first method, only those of some Bembecia species according to the second. The preparation of the female genitalia needs no special comments.

IMMATURE STAGES

Several groups of lepidopterous larvae have now been subjected to taxonomic studies employing detailed descriptions of chaetotaxy (bristle arrangement) of the head and of the trunk segments. MacKay (1968) has treated the available material of Nearctic Sesiid larvae along these lines, but no comparable work has so far been done on European species. Kemner (1922) published a key to the majority of Scandinavian Sesiid larvae, based on a limited number of rather easily observable characters, particularly the shape of the frontoclypeal plate. In our experience, however, the actual variation of the characters in question cannot be adequately described in terms of the clear-cut character states utilized by Kemner; this is particularly evident when some additional species are considered. Consequently we have decided against providing a structural key to the larvae.

Like most endophagous (i.e., feeding inside plant tissues) larvae, those of Sesiidae are devoid of any pigmented pattern (Fig. 8). The head-capsule and pronotal shield are sclerotized and brown; the trunk segments are otherwise whitish. The prolegs are well developed, with crotchets usually arranged uniserially in two transverse rows; in Pennisetia the prolegs on segment VI are reduced. The dorsal area of the composite abdominal segments X-XI, the anal shield, in some species may bear one or two sclerotized spines, but the presence of these is less constant than supposed by Kemner (1922).

Empty Sesiid pupal skins, protruding from emergence holes in the host plants, are familiar objects to field lepidopterists, and occasionally faunistic records have been based solely on such skins. Kemner (1922) also gave a structural key to the majority of Scandinavian Sesiid pupae, and we have been able to examine empty pupal skins of most of the remaining species. In Kemner's key extensive use is made of the lengths of some appendages relative to others, but it must be noted that these relationships are difficult to evaluate in empty skins where some distortion has invariably taken place. We have found it impossible at the moment to provide a structural key which permits a satisfactory identification of pupal skins of the known Scandinavian Sesiids.

Sesiid pupae (Fig. 9) are incomplete, i.e., they have several movable abdominal segments: III-VI (-VII in males). The frontal sclerite of the head usually

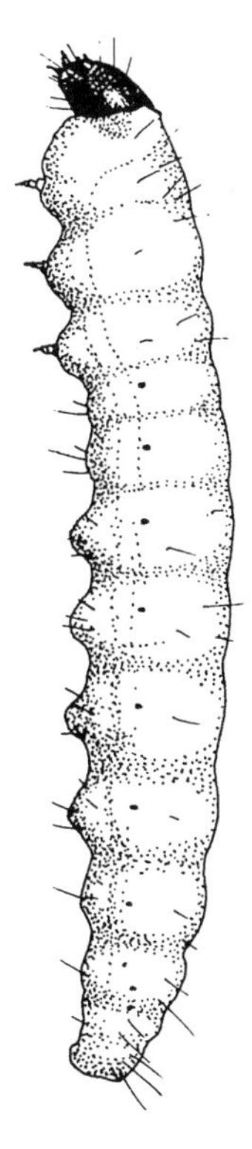

Fig. 8. Larva of _Sesia apiformis_ (Cl.).

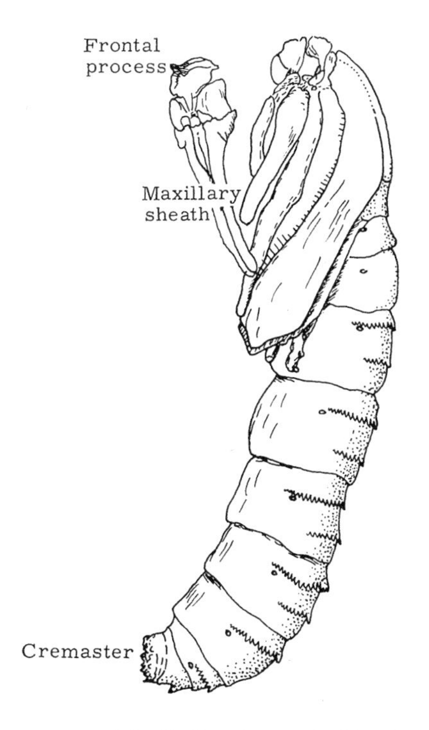

Fig. 9. Empty pupal skin of _Aegeria scoliaeformis_ (Bkh.). (redrawn after Kusnezov).

has a more or less distinctly produced part which is used for opening the cocoon and/or the lid of the emergence hole; the shape of this *frontal process* is taxonomically useful. Dorsally the frontal process merges into the *vertex*, which usually is more or less depressed on each side of the midline. Abdominal segments II-IX are equipped with posteriorly directed spines arranged in transverse rows. A single or (more rarely) two such rows are present on segment II, two on segments III-VI (-VII in males) and one on segments VII- (VIII- in males) IX; the number of spinerows on segment VII thus provides a convenient method of sexing Sesiid pupae (Kemner, 1922).

The structure of the *cremaster*, i.e., the genito-anal region with surrounding stout spines, seems of little usefulness in this family due to considerable intraspecific variation.

In most cases immature stages of Sesiidae can be satisfactorily identified on the basis of knowledge of the food plant and mode of life. Determinations of pupae of wood-boring Aegeriini should be checked by examination of the frontal process.

Bionomics

Adult Sesiidae normally emerge in early morning. Wing expansion is accomplished in a very short time and males may immediately afterwards start a rapid, irregular flight in search of females, which apparently often fly only after copulation. All species are diurnal and dependent on sunshine for activity; the occasional specimens attracted by collecting lights at night have probably always been accidentally disturbed. A few species have the proboscis nonfunctional, but the majority will occasionally visit flowers to feed on nectar; some also suck the exuding sap of wounded trees. In any case the duration of adult life seems to be short, rarely exceeding one week.

The mimetic nature of Sesiid resemblance to Hymenoptera is now probably generally accepted. Some species have been reported to exhibit wasplike postures and movements (Naumann, 1971).

The eggs are often laid in crevices on the host plant surface, whereby the penetration of the newly hatched larva is facilitated. In some cases, however, females have been observed simply dropping the eggs on the ground near the host plant (Kemner, 1922).

Sesiid larvae are internal borers in woody or herbaceous plants. Those of the last-mentioned category are probably generally annual, whereas many of the wood-borers hibernate twice or more. Unless otherwise stated, the life-cycles described are those characteristic of S. Scandinavian populations; at higher latitudes larval life may be prolonged. Pupation probably always takes place in the spring immediately preceding the adult period. Since individual variation in growth rate may be considerable, 1st-and 2nd-year larvae cannot always be separated on size alone, but apparently the tunnels of 1st-year larvae always are ellipsoidal in cross section whereas those of 2nd-year larvae are circular. Wood-boring Sesiid larvae are presumably predominantly sap-feeders (Kemner, 1922). Before pupation the larva constructs the future emergence-hole in the trunk or branch surface. The hole is usually covered with a delicate lid, which can be opened by the emerging pharate adult by means of the specialized frontal sclerite of the pupal cuticle; very rarely is the hole uncovered. Pupation may take place with or without the formation of a cocoon, and even rather closely related species may differ in this respect. The cocoon is externally covered with wood scrapings or frass. Herb-boring species may pupate in the stems of their host plants, or pupation may take place in the earth, inside a silken tube which leads to the surface and through which the pharate adult may easily move upwards.

TECHNICAL NOTES. Successful rearing of most Sesiids demands collecting of near-mature larvae or of pupae. Obtaining larvae or pupae of wood-boring species always involves some removal of branches or cutting into trunks and stumps; care should be taken, of course, to minimize damage to living trees and bushes. If the cocoon of a mature larva is broken in removal, it may be renewed if the larva is placed in a container with sawdust. Branches or blocks of wood containing larvae or pupae should be placed in moist sand; they may be exposed to gentle heat, but never to direct sunlight. Larvae kept warm in younger instars generally yield small adults. It is important to keep humidity sufficiently high by sprinkling; drying may cause shrinkage of wood pieces or cocoons whereby the animals are crushed. Pupae generally appear rather insensitive to mechanical disturbances.

It is often possible to artificially shorten the dormancy of larvae collected immediately before the last hibernation. The larvae may be chilled for a few days in a freezing cabinet; if subsequently kept at room temperature they will soon pupate and the adults emerge within a couple of weeks.

Distribution

As previously stated, the faunistics of Scandinavian Sesiidae are still rather inadequately known; in particular it must be stressed that since some species lend themselves more easily to observation than others, differences in distribution patterns may in some cases be more apparent than real.

About half of the Scandinavian species (Pennisetia hylaeiformis; Sesia apiformis and S. bembeciformis; Aegeria scoliaeformis, A. spheciformis, A. culiciformis, A. formicaeformis and A. tipuliformis) occur in all four countries and range northwards beyond the arctic circle. Two other species, S. melanocephala and Paranthrene tabaniformis, do not extend as far north and are apparently most common in the eastern part of the area although both may have been largely overlooked. One species, A. flaviventris, reaches only the S. E. and E. periphery of the area, i.e., Bornholm and S. and S. E. Finland. Among the southernly distributed species, two (A. myopaeformis and A. vespiformis) seem to be extremely rare and peculiarly scattered in their Scandinavian distribution, while two others are fairly evenly distributed just to their northern limit, which may encircle the southern parts of all four countries (Bembecia scopigera) or Denmark and S. Sweden only (B. muscaeformis). All of the above-mentioned species are widely distributed in N. Europe, although P. hylaeiformis and S. melanocephala are absent from the Netherlands and the British Isles. This is particularly noteworthy in the case of the first-mentioned species, which is one of the most common Scandinavian Sesiids. Two species have an isolated occurrence in Scandinavia: A. polaris, which is restricted to the arctic and subarctic region of N. Europe, and A. mesiaeformis, a truly continental species the westernmost populations of which are markedly disjunct: S. Finland and scattered areas in S. E. Europe.

It is noteworthy that the species-diversity among Sesiids associated with herbaceous plants diminishes much more rapidly towards north in Europe than does that of the wood-borers. Thus, among the 43 C. European species of Sesiidae listed by Forster & Wohlfahrt (1960) 23 belong to herb-feeding groups, whereas of the 17 Scandinavian species only two are associated with herbs.

Classification

The Sesiidae is an almost cosmopolitan family with about 1000 species now described (Naumann, 1971). No world-embracing higher classification exists, but Naumann's important study of the Holarctic genera has provided a classificatory framework into which a great part of the world fauna may probably be fitted. Other recent classifications (Niculescu, 1964; MacKay, 1968) are critically reviewed by Naumann and need no further discussion here.

According to Naumann the Holarctic Sesiid genera fall into two groups, which are given subfamily rank: Tinthiinae and Sesiinae. The former is by far the smallest, and since it is largely defined by the lack of specializations present in Sesiinae, it is probably paraphyletic. The absence of a gnathos in the Tinthiinae might, however, be considered a derived feature, since a gnathos is present in many other lepidopterous families. Two tribes are recognized within the Tinthiinae, one of which, Pennisetiini, is represented in Scandinavia.

The Sesiinae are characterized by clavate antennae with apical hair-pencil, lack of HW A_1, presence of processus vinculi and a trend towards specialization of the valve sensillae. Apart from a primitive and isolated nearctic genus (Calasesia Beutenmüller, 1899), the members of this subfamily fall into four tribes: Sesiini, Paranthrenini, Melittiini (extra-European) and Aegeriini. The Sesiini, primarily characterized by a posterior shift in the FW radial fork, appears to be the primitive sister-group of the three remaining tribes. These tribes (the "higher Sesiinae") share specializations of the anal tufts, elongation and apical curvature of the anellus, and absence of a phallic caecum; each tribe is characterized by specializations in the genitalia.

Nomenclature

The generic nomenclature of European Sesiidae has been extremely confused because of erroneous concepts concerning type-species designations. For a considerable period the generic name Sesia was attributed to a Sphingid group, the name Aegeria being applied to the true Sesia and the family consequently being known as Aegeriidae. The formal problems concerning Sesiid generic nomenclature may now be considered largely solved through Naumann's painstaking examination of all genus-group names in the family, but the delimitation of individual genera will, of course, continue to vary according to subjective judgements of subsequent authors.

The specific names of N.W. European Sesiidae have remained relatively stable throughout this century. Two alterations in specific names have very recently been introduced in the British checklist by Bradley et al. (1972), but application is being made to the International Commission on Zoological Nomenclature for protection of the names used hitherto (Kristensen, in press a & b).

Synonyms are mentioned in this work only to the extent necessitated for continuity with the comprehensive Scandinavian works listed above and with Edelsten et al.'s (1961) treatment of the British species.

Key to subfamilies, tribes and genera

1 Antennae without terminal hair-pencil. FW PTA basal to ATA. Facies as in Figs. 83-84 (Tinthiinae, Pennisetiini)... .... Pennisetia Dehne (p. 25)

- Antennae with apical hair-pencil. FW PTA at least partially behind ATA (Sesiinae).. 2

2 (1) FW largely covered with pigmented scales, ETA absent, ATA and PTA indicated basally only (Fig. 98) (Paranthrenini) .... Paranthrene Hübner (p. 34)

- FW largely hyaline, at least ATA and ETA well developed 3

3 (2) Large and robust forms. FW R4 to apex, R5 to outer margin. Apical area diffusely delimited or entirely lacking. Discal spot not continued to hind margin (Sesiini) .. Sesia Fabr. (p. 27)

- Slender forms. FW R5 to apex. Apical area large, well defined. Discal spot usually more or less distinctly continued to hind margin along Cu2 (Aegeriini) 4

4 (3) FW PTA long, reaching discal spot .... Aegeria Fabr. (p. 38)

- FW PTA short, not reaching discal spot or absent 5

5 (4) Male genitalia: Scopula androconalis present, valvae without crest delimiting sensory field. Facies as Figs. 105-110 ... . Bembecia Hübner (p. 63)

- Male genitalia: Scopula androconalis absent, valvae with crest delimiting sensory field. NO SPECIES YET RECORDED FROM SCANDINAVIA OR BRITISH ISLES .... Chamaesphecia Spuler (appendix p. 81)

Survey of immature stages based on the typical food plants

CONIFERAE

Juniperus .... Aegeria spuleri (Fuchs)

BETULACEAE

Betula and Alnus, often in stumps.

Old larvae always in the wood. No cocoon ....
.... 8. Aegeria spheciformis (Den. & Schiff.)

Larvae sometimes remaining in boundary layer between bark and wood. Scrapings highly characteristic, long chips. Cocoon present .... 9. Aegeria culiciformis (L.)

Betula

Between bark and wood of old trunks. Cocoon present........... .... 6. Aegeria scoliaeformis (Bkh.)

Alnus

Between bark and wood of old trunks. Cocoon present .... 7. Aegeria mesiaeformis (H.S.)

CUPULIFERAE

Quercus

Diameter of tunnel distinctly larger than that of larva .... Aegeria conopiformis (Esp.)

Diameter of tunnel just matching that of larva... 15. Aegeria vespiformis (L.)

SALICACEAE

Salix and Populus, ground-near parts of larger trunks. Cocoon immediately beneath emergence hole .... 3. Sesia apiformis (Cl.)

Salix

S. caprea, ground-near parts of trunk. Emergence hole uncovered (a unique condition), far from terminal part of tunnel with cocoon .... 4. Sesia bembeciformis (Hübn.)

S. lapponum (more rarely S. caprea). Frass extruded. Emergence hole a few centimeters above ground level. No cocoon .... 12. Aegeria polaris (Stgr.)

S. repens and S. aurita, swellings formed on twigs; leaves beyond swelling pale. No cocoon .... 14. Aegeria flaviventris (Stgr.)

Several species of Salix, frass extruded. Swellings formed on branches of S. repens and S. caprea. No cocoon .... 11. Aegeria formicaeformis (Esp.)

Populus

P. tremula. In dead branches or adjacent parts of trunk. No cocoon .... 2. Sesia melanocephala Dalm.

Several species of Populus. In galls on branches or in lower parts of trunks. No cocoon .... 5. Paranthrene tabaniformis (Rott.)

POLYGONACEAE

Rumex

R. acetosa and R. crispus .... Bembecia chrysidiformis (Esp.)
R. acetosella .... Bembecia triannuliformis (Frr.)

RIBESIACEAE

Ribes .... 13. Aegeria tipuliformis (Cl.)

POMACEAE, AMYGDALACEAE

Several genera .... 10. Aegeria myopaeformis (Bkh.)

ROSACEAE

Rubus .... 1. Pennisetia hylaeiformis (Lasp.)

PAPILIONACEAE

Several genera, notably Anthyllis, Ononis, Lotus ....
.... 16. Bembecia scopigera (Scop.)

PLUMBAGINACEAE

Armeria .... 17. Bembecia muscaeformis (Esp.)

CAPRIFOLIACEAE

Viburnum .... Aegeria andrenaeformis (Lasp.)

It must be stressed that only typical food-plants are listed in this survey. Information on secondary hosts are given in the treatment of each species.

SUBFAMILY TINTHIINAE

As previously stated this subfamily includes the most overall primitive Holarctic Sesiids and is probably a paraphyletic taxon. One tribe represented in Scandinavia.

Tribus Pennisetiini

Characterized by the absence of FW M_3, by FW R_4 and R_5 being stalked or coincident and by a specialized ductus bursae. One Scandinavian genus and species.

Genus *Pennisetia* Dehne, 1850

Pennisetia Dehne, 1850, Stettin ent. Ztg., 11: 28.
  Type-species: Pennisetia anomala Dehne, 1850 = Sesia hylaeiformis Laspeyres, 1801.
Bembecia auct.

FW R_4 and R_5 as well as HW M_3 and Cu_1 stalked. Male antennae markedly bipectinate.

1. PENNISETIA HYLAEIFORMIS (Laspeyres, 1801)
   Figs. 10-11, 83-84, 111-112.

Sesia hylaeiformis Laspeyres, 1801: 14 n.7.

A robust species distinctive by the narrow FW with relatively small transparent areas, PTA entirely basal to ATA. ETA transversed by only two veins. Abdomen dorsally with broad yellow bands posteriorly on IV-VI (-VII, males only) and very narrow bands posteriorly on III as well as anteriorly on II-III; the narrow bands are frequently virtually invisible, as in the illustrated female specimen. The anal tufts of both sexes may be suffused with yellow or orange to a variable extent. Proboscis short, weakly sclerotized. FW 9-14 mm. - Male genitalia (Fig. 10): Gnathos lacking, valvae simple, anellus almost tubular with posterior processes. - Female genitalia (Fig. 11): Venter VIII with distinct lamella postvaginalis. Ductus bursae looped, proximally sclerotized; antrum somewhat widened, its lateral outline smoothly convex.

Larva without functional prolegs on VI, at most with a few isolated crotchet-vestiges on low protuberances in the position of the prolegs.

Pupa with short maxillae, single row of spines on II and triangular frontal process.

Common in all Scandinavian countries, ranging northwards beyond the Arctic circle. - Widely distributed in the Palaearctic region, but absent from the Netherlands and the British Isles.

The adults emerge during a long period, from the middle of June until the beginning of August. The moths frequently rest on blossoms or foliage of the food plant. As noted by Hoffmeyer (1960), previous reports of regular nocturnal activity of this species seem erroneous. The food plant is usually Rubus idaeus, particularly cultivated forms, less typically R. fruticosus. The eggs are laid on ground-near parts of stems or rootstocks or simply dropped on the earth (Kemner, 1922). The annual larva tunnels in the rootstock. Before hibernation, which takes place in the rootstock, it makes a tunnel into the pith of the stem. In spring this tunnel is extended further into the stem (which at this time is dead), where pupation takes place without formation of a cocoon (Figs. 111-112). The emergence hole, covered by a thin lid, may be located 0-25 cm above ground level. On rare occasions larvae and pupae are found in sound stems.

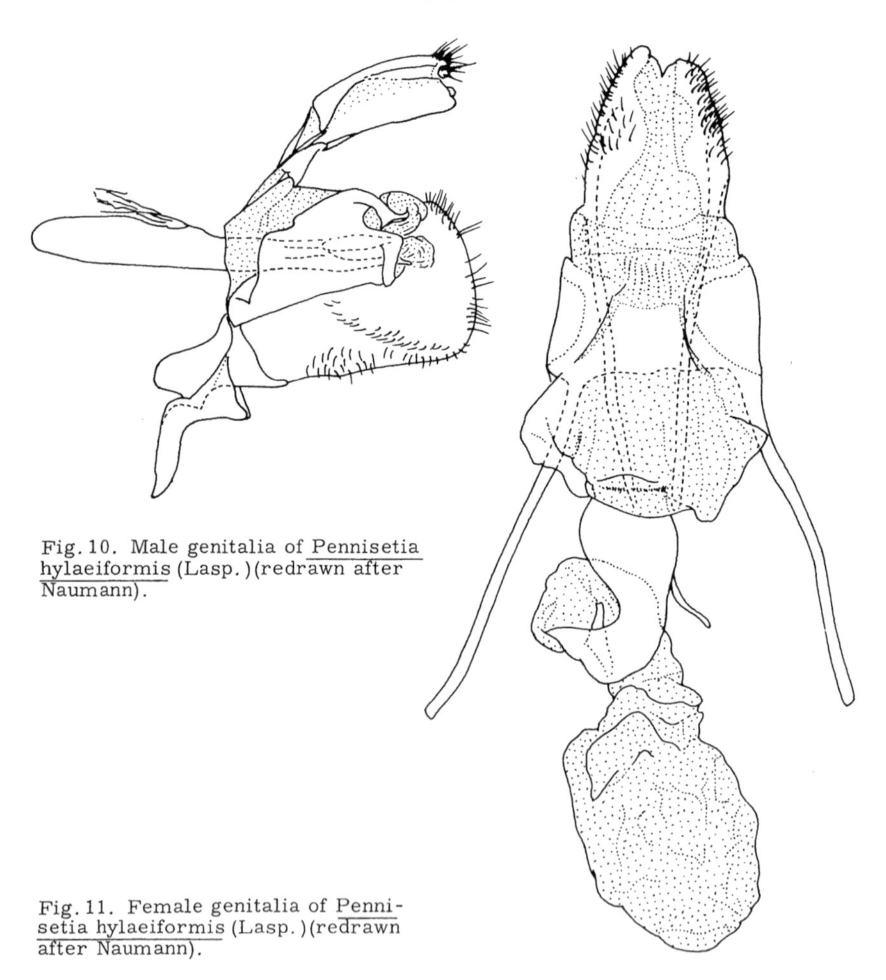

Fig. 10. Male genitalia of Pennisetia hylaeiformis (Lasp.)(redrawn after Naumann).

Fig. 11. Female genitalia of Pennisetia hylaeiformis (Lasp.)(redrawn after Naumann).

Preferred habitats are gardens and plantations where the species may be noxious; furthermore wood glades and thickets in moist localities.

Specimens are obtained by cutting, in June, stems of food plant individuals from the preceding year just above the rootstock. If frass is not found in infested stems, one should check to see whether the larva is still in the rootstock.

SUBFAMILY SESIINAE

This subfamily represents an overall higher level of organization than the Tinthiinae, and is much richer in species. Three tribes are represented in Scandinavia, see key p. 22.

Tribus Sesiini

Large, robust Sesiids with FW R_4 running into the wing apex and R_5 to the outer margin. As regards the structure of the male genitalia this is the most overall primitive tribe within the subfamily. One genus in Scandinavia.

Genus *Sesia* Fabricius, 1775

Sesia Fabricius, 1775, Syst. ent.: 547.

Type-species: Sphinx apiformis Linné, 1761 = Sphinx apiformis Clerck, 1759.

Sphecia Hübner, 1819, Verz. bek. Schmett.: 127.

Type-species: Sphinx crabroniformis Lewin, 1797 (nec Denis & Schiffermüller, 1775) = Sphinx bembeciformis Hübner, 1787.

Eusphecia Le Cerf, 1937, Z. öst. ent. Ver., 22: 106, syn. n.

Type-species: Sesia pimplaeformis (Boisduval in litt.) Oberthür, 1872.

Characterized by the presence of thickenings or processes on the valvae (Naumann, 1971) and by the anteriorly bifid phallic coecum.

Note. S. bembeciformis is often separated from S. apiformis at the generic level, but as pointed out by Popescu-Gorj et al. (1958), Niculescu (1962), Schwarz & Niculescu (1962) and Naumann (1971), this distinction is untenable. The similarities in facies and genitalia are considerable, and some of the previously described differences in the mouthparts (proboscis development and palp structure) are virtually nonexistent.

Naumann (1971:119) predicted the existence of intermediate forms between the genera Sesia and Eusphecia. Sesia melanocephala is actually such an intermediate type. Like the S.W. Asian species pimplaeformis (type-species of Eusphecia), melanocephala has an apically pointed valve (Fig. 13), medially carrying a conspicuous ventral process and a dense dorsal covering of long hairs, which (particularly on the basal part of the valve) are widened and cleft apically. Also the phallus and the dorsally conspicuously spined manica are very similar in the two species (Fig. 12). On the other hand, melanocephala like apiformis (type species of Sesia) and bembeciformis has retained well-developed gnathos lobes, whereas these are much shortened in pimplaeformis. In the corpus bursae of melanocephala (Fig. 14) a signum is present as in apiformis-bembeciformis, while, as in pimplaeformis, branches of the anterior apophyses support the lamella antevaginalis, and the antrum has a distinct, coherent sclerotization. There can be little doubt that melanocephala is phylogenetically close to pimplaeformis and it might have been included in Eusphecia, but considering the diminishing phenetic gap between the Sesia and Eusphecia lineages it seems preferable to treat the members of both as constituting a single genus.

Key to the species of Sesia

1 Abdomen dorsally with relatively narrow yellow bands anteriorly on II-IV and posteriorly on V-VI (-VII in males) (Figs. 85-86) .... 2. melanocephala Dalm.

\- Abdominal segments beyond II and except IV usually predominantly yellow, posteriorly and medially with dark suffusion of variable extent 2

2 (1) Patagial collar dark, tegulae anteriorly with prominent yellow patch (Fig. 87) .... 3. apiformis Clerck

\- Patagial collar bright yellow, tegulae unicolorous dark (Fig. 89) ... . 4. bembeciformis (Hbn)

2. SESIA MELANOCEPHALA Dalman, 1816

Figs. 12-14, 85-86, 113-115.

Sesia melanocephala Dalman, 1816: 217.

Immediately distinctive by the abdominal colour-pattern as stated in the key. Male antennae more markedly bipectinate than in the other Sesia species and FW pigmented scales permanent to greater extent. Anal tufts in female more or less suffused yellow. FW 12-15 mm. - Genitalia (Figs. 12-14): See above.

Pupal maxillary sheath basally with angular process, less prominent than in S. apiformis.

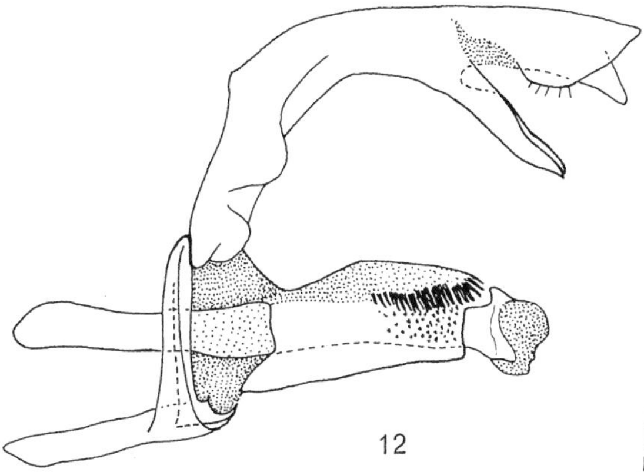

12

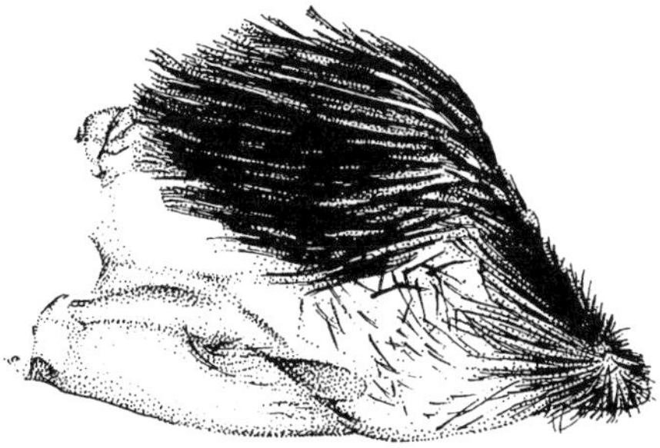

13

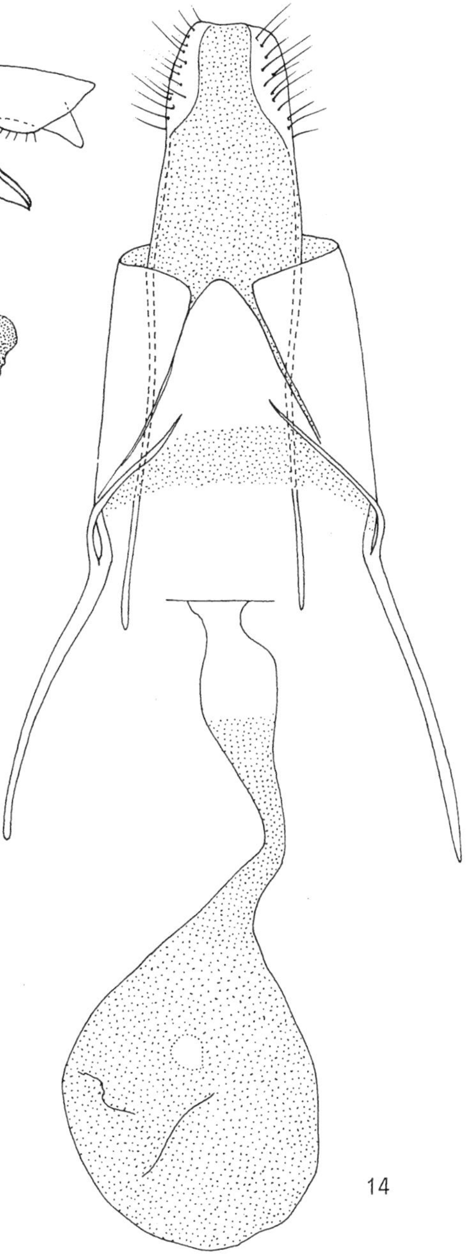

14

Figs. 12-14. Genitalia of Sesia melanocephala Dalm. - 12: male genital segments; 13: male valve; 14: female genitalia.

Recorded from all Scandinavian countries. In Denmark only discovered in the 1960's, now known from scattered localities in E. Jutland and N. E. Zealand. In Norway only one record (Akershus). Several records from Sweden (particularly E. provinces) and Finland, northernmost in Ångermanland and Ostrobottnia media respectively. - Very scattered records from W. Palaearctic region; apparently always local and rare. Absent from the Netherlands and British Isles.

The adults occur from the beginning of June until late July. Emergence starts about 0400 in the morning. Females do not assemble males until the second day after emergence, probably because of retarded development of scent glands (Nordman, 1963); they do not fly before copulation. The species appears to be monophagous on *Populus tremula*. The eggs are laid in the vicinity of dead branches or in bark crevices on the trunk. The young larva bores a tunnel around the base of the dead branch. In autumn it enters the dead knot from where it eventually goes into the central part of the trunk; in this well sheltered place hibernation takes place in a system of short, irregular tunnels. In spring the larva returns to the dead branch where it bores until next autumn, when it constructs the future pupal tunnel before returning to the trunk for the second hibernation. In the following spring it makes the emergence hole, which never faces upwards and which is covered with a thin lid. The terminal portion of the tunnel has a delicate silken lining, but no cocoon is made. Consequently the pupa is capable of rapid backwards movement upon disturbance, which probably to some extent accounts for the apparent low level of woodpecker predation upon this species.

The preferred habitats are groups of old *Populus tremula*, fairly sun-exposed and standing on sandy or rocky soils. The search for *S. melanocephala* is extremely time-consuming. Larvae and pupae are obtained by cutting off and examining dead branches 0.2-6 m above ground level; branches containing a tunnel are kept for rearing. The part of the cavity left inside the trunk should be explored since larvae or pupae located at the transition between trunk and branch will escape by backwards movements. Younger larvae (recognizable by ellipsoidal cross-section of tunnel rather than by size of larvae may be enclosed in artificial chambers bored into the central part of a trunk. The cavity should have a diameter corresponding to that of the larva and should be covered by a wire gauze cage which is inspected every morning during the appropriate emergence period. Pupae should be kept dry.

3. *SESIA APIFORMIS* (Clerck, 1759)

Figs. 15, 17, 87, 116-120.

Sphinx apiformis Clerck, 1759: Pl. 9 fig. 2.

This large, hornet-like species is likely to be confused only with the following, from which it is separable by the coloration of the patagial collar and the tegulae as stated in the key. Facial scales bright yellow. HW M_3 and Cu_1 arising from same point at lower corner of cell or very shortly stalked. FW 15-21 mm. - Male genitalia (Fig. 15): tegumen ventrally excavated. Medially on valve a thickening with long and stout setae; ventro-apical valval corner with row of setae. - Female genitalia (Fig. 17): venter VIII membranous. Antrum with a weak, annular sclerite.

Pupal maxillary sheath basally with distinct angular process.

All Scandinavian countries, common; northernmost record from Ostrobottnia borealis (Finland). - Widely distributed in W. and C. Palaearctic region and accidentally introduced into N. America.

The adults occur in June-July (rarely in August). Emergence takes place about 0500-0800 in the morning. Later in the morning copulating animals are often observed on the trunks of the host trees. Oviposition takes place in the afternoon, the eggs being placed at low heights in bark crevices or in old emergence holes. The host plant may be several species of Populus and Salix, but Populus nigra seems to be preferred. The larva makes irregular tunnels (Fig.

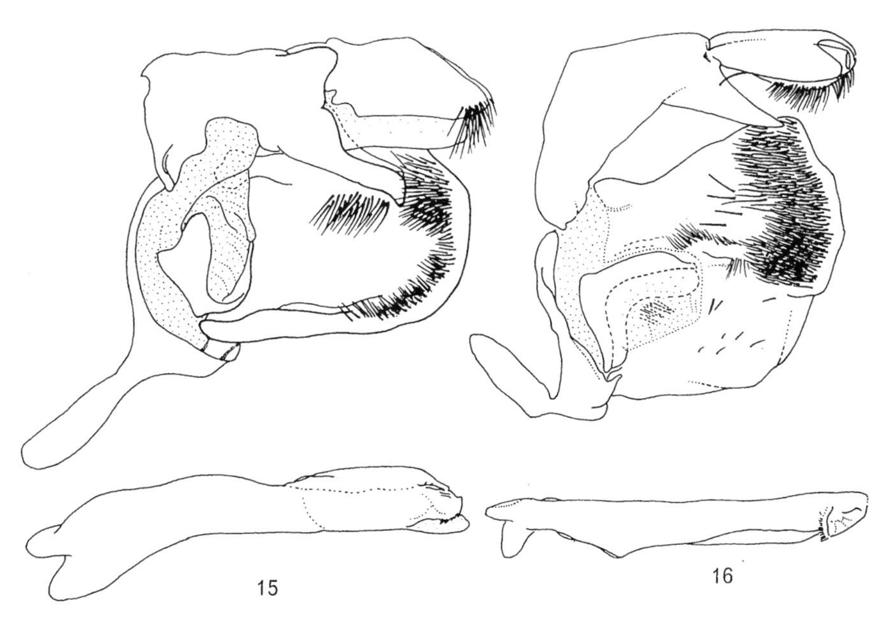

Figs. 15-16. Male genitalia of Sesia. - 15: S. apiformis (Cl.); 16: S. bembeciformis (Hübn.). (redrawn after Naumann).

117) between bark and wood, keeping near ground level (often going into the roots). In heavily infested trees the larvae may bore deeper into the wood. The frass is extruded from the tunnels and may form great heaps around the tree. The larva hibernates twice. Before the second hibernation it makes a dense cocoon, covered with wood scrapings. Pupation takes place in April or May. The cocoons are preferably placed in the sunny side of the trunk and at low heights, hidden by the surrounding grass; they are covered by a thin bark lid (Figs. 118-120). When such suitable pupation sites are lacking, particularly in heavily infested trees, the cocoon may be formed with earth-particles in the ground outside the tree, or the larva may ascend the trunk through cavities below the bark and pupate at greater heights (up to 1 meter above ground level).

Preferred habitats are avenues, edges of woods, etc.

S. apiformis is most easily obtained as a mature larva by removing bark from infested trees; the covered emergence holes cannot normally be revealed by brushing the bark.

4. SESIA BEMBECIFORMIS (Hübner, 1797)
Figs. 16, 18, 89, 121-123.

Sphinx bembeciformis Hübner, 1797: 92.
Sphinx crabroniformis Lewin, 1797: 2 (nec Denis & Schiffermüller, 1775).

Distinguished from S. apiformis by the dark face (occasionally with dull yellow suffusion), brightly yellow patagial collar and unicolorous tegulae. Dark suffusion on abdomen generally less extensive. Wings with extremely few pigmented scales. HW M_3 and Cu_1 usually much more distinctly stalked. FW 14-18 mm. - Male genitalia (Fig. 16): tegumen without ventral concavity. Thickening medially on valve with short and fine setae only; ventro-apical valval corner without setae. - Female genitalia (Fig. 18): sclerotized lamella postvaginalis present. Lateral lips of ostium weakly sclerotized. Antrum with three pairs of semicircular sclerotizations. Pupal maxillary sheath smooth.

Scattered records from Denmark. Sweden: Skåne. Norway: Sør-Trøndelag, Garten, 1971 (S. A. Aas leg., M. Opheim verif. 1974). Finland: Savonia borealis and as far north as Lapponia kemensis. Probably largely overlooked. - N. W. and C. Europe.

The adults occur in June-July. After copulation the females fly actively and oviposit until about 1500. The eggs are laid close to the ground on trunks of Salix caprea (rarely S. viminalis or S. alba), often several eggs per tree. The young larvae bore irregular tunnels below the bark, often descending below ground level, but indicating its presence by extrusion of the finely granular frass. After the first hibernation, which takes place below ground level, the larvae will commence boring into the wood, occluding the 1st-year tunnels with coarse frass. In autumn the emergence hole, 0.7-1 cm wide, is made about 10 cm above ground level. The hole usually faces upwards (Fig. 121) and is without any cover, contrary to conditions in other wood-boring Sesiids. Eventually the

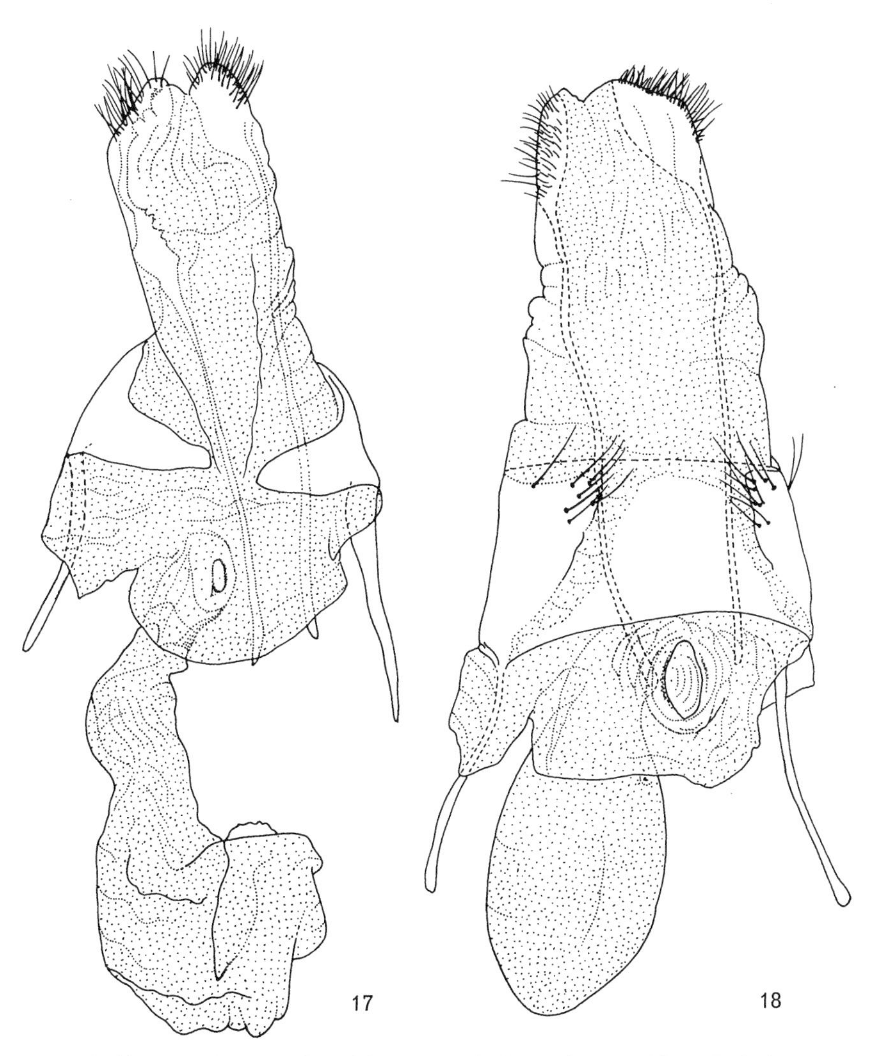

Figs. 17-18. Female genitalia of Sesia. - 17: S. apiformis (Cl.); 18: S. bembeciformis (Hübn.)(redrawn after Naumann).

cocoon is constructed in the uppermost end of the tunnel (Figs. 122-123), usually 15-25 (but occasionally about 50) cm above the emergence hole. The upper portion of the cocoon is a delicate silken lining of the tunnel wall, whereas the down-facing part is tough, mixed with wood scrapings. Pupation takes place in spring.

Preferred habitats are moderately open groups of Salix caprea in moist localities; sometimes the species is encountered also in hedges on dryer soil. The larvae and pupae are heavily preyed upon by woodpeckers.

In early winter (before peak of woodpecker predation) the emergence holes may be disclosed by removing the low vegetation covering the lower parts of the trunks. Old tunnels are generally darker than new, inhabited ones. The length of the upper tunnel may be measured (e.g., by means of a straw gently inserted into the hole) and a block of wood containing the cocoon subsequently carefully cut out. Alternatively the entire trunk may be cut down and a c. 50 cm long piece above the hole cut free. As a third method a metal gauze cage may be fitted around the hole and inspected every morning in the emergence period.

Tribus Paranthrenini

This tribe is characterized (Naumann, 1971) by the uncus being elongated with concomitant reduction of the tegumen, by the usual presence of more or less distinct subapical hooks on the phallus, by the uncus being dorsally scale-covered and laterally hairy. The female anterior apophysis is with posteriorly directed, ventral branch. One genus and species in Scandinavia.

Genus *Paranthrene* Hübner, 1819

Paranthrene Hübner, 1819, Verz. bek. Schmett.: 128.

Type-species: Sphinx vespiformis Linné sensu Newmann, 1832 = Sphinx tabaniformis Rottemburg, 1775.

Sciapteron Staudinger, 1854, De Sesiis Agri Berol.: 39, 43.

Type-species: Sphinx asiliformis (Denis & Schiffermüller, 1775). = Sphinx tabaniformis Rottemburg, 1775.

This genus belongs to a section within the tribe characterized by the tegumen being indistinguishably fused with the uncus, the valve having specialized dorsal setae and the corpus bursae having a most distinctly ribbed wall.

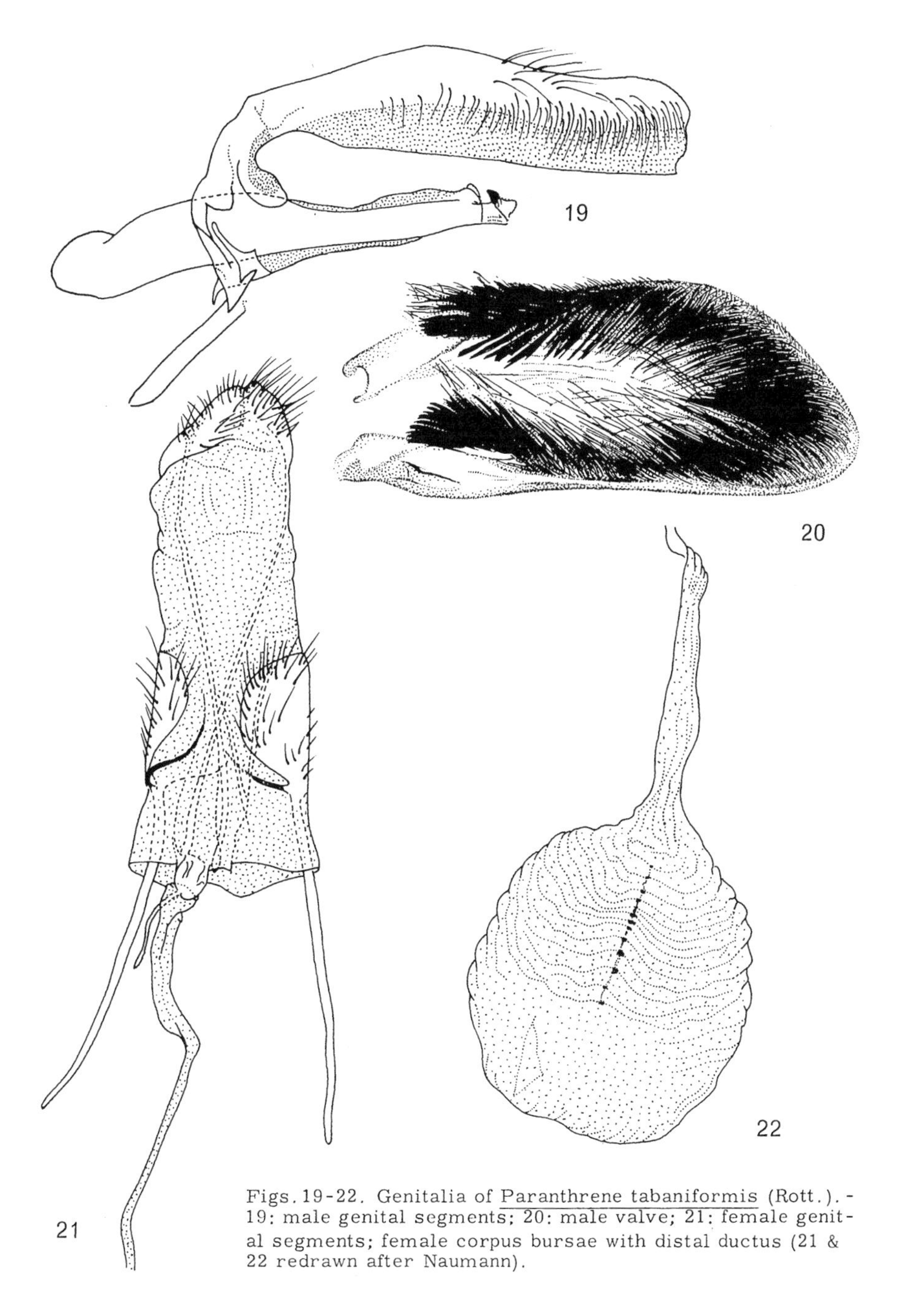

Figs. 19-22. Genitalia of Paranthrene tabaniformis (Rott.). - 19: male genital segments; 20: male valve; 21: female genital segments; female corpus bursae with distal ductus (21 & 22 redrawn after Naumann).

5. PARANTHRENE TABANIFORMIS (Rottemburg, 1775)
Figs. 19-22, 88, 124-126.

Sphinx tabaniformis Rottemburg, 1775: 110.

The only N.W. European Sesiid with FW almost entirely covered with pigmented scales; only near the base are the ATA and PTA indicated to a variable extent. Abdomen dorsally with yellow transverse bands posteriorly on II, IV, VI and (in males) VII. Antennae in males markedly bipectinate. FW 11-14 mm. - Genitalia (Figs. 19-22): see above.

Pupal maxillary sheaths smooth.

In Denmark known from Jutland and Bornholm. In Sweden recorded from Västergötland and middle E. provinces to Hälsingland. In Norway only old records from S.E. provinces. In S. Finland, northernmost in Karelia borealis. Probably largely overlooked. - W. and C. Palaearctic region; some S. European and N. African populations may represent distinct species (Popescu-Gorj & Capuse, 1969) but this needs further clarification.

The adults occur from the end of May until the middle of July. The moths rest on twigs or foliage of the food plants, which usually are Populus tremula, P. nigra or P. canadensis, very occasionally Salix species. The eggs are laid in the afternoon, and according to mode of oviposition, larval development may follow one of two different patterns (Urbahn, 1939):

1). The eggs are placed on the bases of leaves on twigs or suckers. The newly emerged larva lives in a web outside the bark, and hibernation takes place inside a gall made by the Cerambycid beetle Saperda populnea. In the spring the larva bores a tunnel, thereby producing a gall (Figs. 124-126), which is distinguishable from the Saperda gall in being pyriform or ellipsoidal instead of globular. The second winter is spent in the gall, and pupation takes place in May.

2) The eggs are placed in bark crevices on trunks near ground level. After hatching the larva bores in the roots or inside the bark. After the first hibernation it makes tunnels in the wood or between bark and wood, and after the second hibernation it pupates in a chamber below the trunk surface, covered only by a very thin layer of bark.

P. tabaniformis does not make a cocoon. The first-mentioned developmental type is probably the most usual, but larvae developing according to the second type are less easily discovered. In C. and S. Europe the damage of this species to poplars may be of considerable economic significance, but heavy attacks are unlikely to occur in Scandinavia.

The species is most easily found by searching for the characteristically shaped gall on suckers or on finger-wide, young trees of Populus tremula.

Tribus Aegeriini

This tribe, which comprises the vast majority of Holarctic Sesiidae, is characterized by specialization of the male genitalia (see p.13 and Figs.6, 23-56), viz. the structure and arrangement of sensillae of the valval sensory field as well as presence of a variously developed crista gnathi. Moreover, most Aegeriini have a scopula androconalis. The Aegeriini are small to medium-sized, slenderly built Sesiids. All Scandinavian species have at least the FW ATA fully developed. The male antennae never have large processes.

The larvae have the full number of prolegs and are devoid of sclerotized spines on the anal shield. The pupae are characterized by the long maxillae.

There is no general agreement on generic delimitation in the Aegeriini. Following the classification in Spuler (1910), the European species have customarily been grouped into three genera on the basis of venational details, extent of PTA and development of proboscis; however, it has become evident that Spuler's generalizations concerning venational patterns are not tenable (Naumann, 1971; Kristensen, unpublished). On the basis of his survey of all Holarctic genera, Naumann (1971) recognized four phenetically distinguishable genus-groups. One of these, the Chamaesphecia-group, is characterized by absence of scopula androconalis, by the valval sensory field being delimited by a crest and by the distal confluence of FW R_1 and R_2. The others, viz. the Alcathoe-group, the Podosesia-group (Nearctic) and the Aegeria-group, all possess a scopula and lack the crest delimiting the sensory field; R_1 and R_2 may be parallel, distally approximating or confluent. The three last-mentioned groups might together constitute the sister-group of the Chamaesphecia-group, but as pointed out by Naumann, the possibility exists that the lack of the scopula in the latter is secondary, since in one of its member-genera (the E. European Weismaniola Naumann, 1971) the lateral hairs of the uncus are of the specialized type occurring on the valve. However this may be, it must be noted that the large genus Chamaesphecia as conceived in the standard handbooks (e.g., Spuler, 1910; Bartel, 1912; Hering, 1932; Forster & Wohlfahrt, 1960) is phenetically heterogenous, including some species-groups which on the genital and venational criteria mentioned above should belong elsewhere in the tribe. One of these groups had a generic name already available, i.e., Pyropteron (type: Sphinx chrysidiformis Esper, 1779) and was by Naumann placed in the Aegeria group.
For two others new subgeneric names, Synansphecia (type: Sesia triannuliformis Freyer, 1843) and Dipchasphecia (type: Dipsosphecia roseiventris Bartel, 1913) were proposed by Capuse (1973), but retaining these groups in the genus

Chamaesphecia very probably renders this taxon paraphyletic in terms of those Aegeriini which have a short PTA and a reduced proboscis.

As expressively stated by Naumann, his Alcathoe-, Podosesia- and Aegeria-groups are very probably not monophyletic, but even their phenetic distinctiveness seems doubtful. The extensively sclerotized ductus bursae (antrum sensu Naumann) characterizing the Alcathoe-group and the process-bearing phallus characterizing the Podosesia-group are both conditions approached by members of the overall less specialized Aegeria-group (see, e.g., the ductus bursae of Aegeria vespiformis, Fig. 69, and the phalli of the two closely related Aegeria species, A. scoliaeformis and A. mesiaeformis, Figs. 23-24).

We shall here tentatively follow Bradley et al. (1972) in including in a single genus, Bembecia, those N. W. European species which possess a scopula androconalis, have a short (or lacking) PTA and whose larvae are associated with herbaceous plants. This genus then includes elements from Naumann's Alcathoe-group (scopigera) and Aegeria-group (chrysidiformis) and from Chamaesphecia s. lat. The remaining N. W. European Aegeriini all belong to Naumann's Aegeria-group and are here included in a single genus, Aegeria (see note below).

Genus *Aegeria* Fabricius, 1807

Aegeria Fabricius, 1807, Illiger's Mag. f. Insektenk., 6: 288.
Type-species: Sphinx culiciformis Linné, 1758.
Conopia Hübner, 1819, Verz. bek. Schmett.: 129.
Type-species: Sphinx stomoxiformis Hübner, 1790.
Synanthedon Hübner, 1819, Verz. bek. Schmett.: 129.
Type-species: Sphinx oestriformis Rottemburg, 1775 = Sphinx vespiformis Linné, 1761.

FW PTA long, reaching discal spot. - Larvae wood-borers.

Note. Naumann (1971) considered the three taxa Aegeria, Conopia and Synanthedon to be valid genera because of the structural (particularly genital) differences between their type-species. It is apparent from his treatment (p. 101) that he would assign the majority of the species in this group to Conopia, whereas the type-species of Aegeria and Synanthedon were considered to be isolated. Bradley et al. (1972) adopted Naumann's partition of the group but included tipuliformis (a species explicitly stated by Naumann, 1971: 136, 140 to belong to Conopia) in the genus Synanthedon. It seems evident, however, that culiciformis (type-species of Aegeria) and vespiformis (type-species of Synanthedon) both have their nearest relatives in the taxon Conopia as conceived by the above-mentioned authors. Thus the distinctive crista sacculi of culiciformis may straightforwardly be considered an extreme specialization of the "high-ridged" type of some other red-banded species, and similarly the peculiar crista gnathi of vespiformis is to some extent approached by that of flaviventris. Because of the overall homogeneous nature of the entire group and of the pronouncedly reticulate character distribution among its species it seems preferable to include them all in a single genus.

Key to species of Aegeria

1	Abdomen with broad red band on segment IV ("Red-banded" group)	11
-	Abdominal bands yellow or whitish, sometimes very indistinct ...	2
2 (1)	Scale covering between eyes uniformly dark	3
-	Laterofacial scales white	5
3 (2)	FW with extensive suffusion of orange-red scales around transparent areas. Abdominal light bands very narrow and easily obliterated (Fig. 97). Northern species .... 12. polaris (Stgr.)	
-	FW without orange-red suffusion	4
4 (3)	Antennae with broad yellowish-white subapical band. Abdomen dorsally with light band on segment II only. FW ETA broad, including basal part of radial fork (Fig. 92) .... 8. spheciformis (Den. & Schiff.)	
-	Antennae unicolorous dark. Abdomen dorsally with light bands on segment II and IV. FW ETA narrow, not including radial fork (Fig. 93)........................... andrenaeformis (Lasp.)	
5 (2)	Anal tuft largely bright orange (Fig. 90) ..... 6. scoliaeformis (Bkh.)	
-	Anal tuft black or black and yellow	6
6 (5)	Antennae with broad yellow subapical band (Fig. 91). Larger species, FW 12 mm. long or longer .... 7. mesiaeformis (H. S.)	
-	Antennae unicolorous dark. Smaller species, FW up to 11 mm. long	7
7 (6)	Hind tibia bright yellow with black markings. FW discal spot externally bright orange-red (Figs. 103-104) .15. vespiformis (L.)	
-	Hind tibia largely dark. FW discal spot uniformly dark or with dull orange suffusion externally	8
8 (7)	Metanotum with yellow transverse band (Fig. 98) . conopiformis (Esp.)	
-	Metanotum uniformly dark	9

9 (8)	Postvertical scales and tegulae dark (Fig. 99). Yellow band on abdominal segment IV ventrally usually expanded into large patch extending to segment VI	14. flaviventris (Stgr.)
-	Postvertical scales and inner tegular margins yellow (Figs. 100-102). No large yellow patch ventrally on abdomen	10
10 (9)	Associated with Ribesiaceae (rarely with Rubus, Corylus, Euonymus)	13. tipuliformis (L.)
-	Associated with Juniperus	spuleri (Fuchs)
11 (1)	FW apical area suffused bright reddish. Anal tuft laterally with whitish streak (Fig. 96)	11. formicaeformis (Esp.)
-	FW apical area as well as anal tuft uniformly dark	12
12 (11)	FW basally suffused red (Fig. 94). Red abdominal band ventrally closed	9. culiciformis (L.)

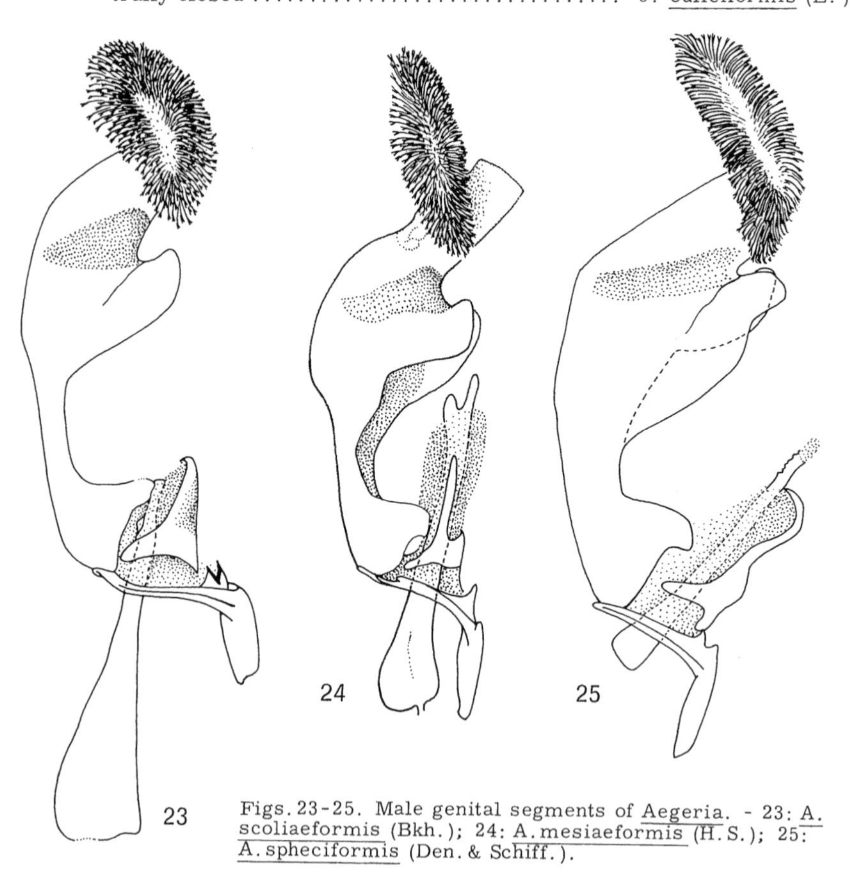

Figs. 23-25. Male genital segments of Aegeria. - 23: A. scoliaeformis (Bkh.); 24: A. mesiaeformis (H.S.); 25: A. spheciformis (Den. & Schiff.).

- FW not basally suffused red (Fig. 95). Red abdominal band not ventrally closed .... 10. myopaeformis (Bkh.)

6. AEGERIA SCOLIAEFORMIS (Borkhausen, 1789)
Figs. 23, 40, 57, 74, 90, 128-130.

Sphinx scoliaeformis Borkhausen, 1789: 173.

Generally the most robust member of the genus. Brilliant orange anal tuft usually most distinctive (not confined to the male as stated by Hoffmeyer, 1960), rarely is the bright colour partly obliterated by dark suffusion. Very broad, inwardly pointed, FW discal spot also distinctive. Female antennae with light subapical band. Laterofacial scales white. Tegulae edged yellow. Abdomen dorsally with yellow bands on segments II and IV. Hind tibia dark with dull yellowish suffusion. FW 12-16 mm. - Male genitalia (Figs. 23, 40): crista sacculi ventral, with narrow scales. Crista gnathi very small. Phallus apically with low dorsal keel. - Female genitalia (Fig. 57): lateral plates of segment VIII in

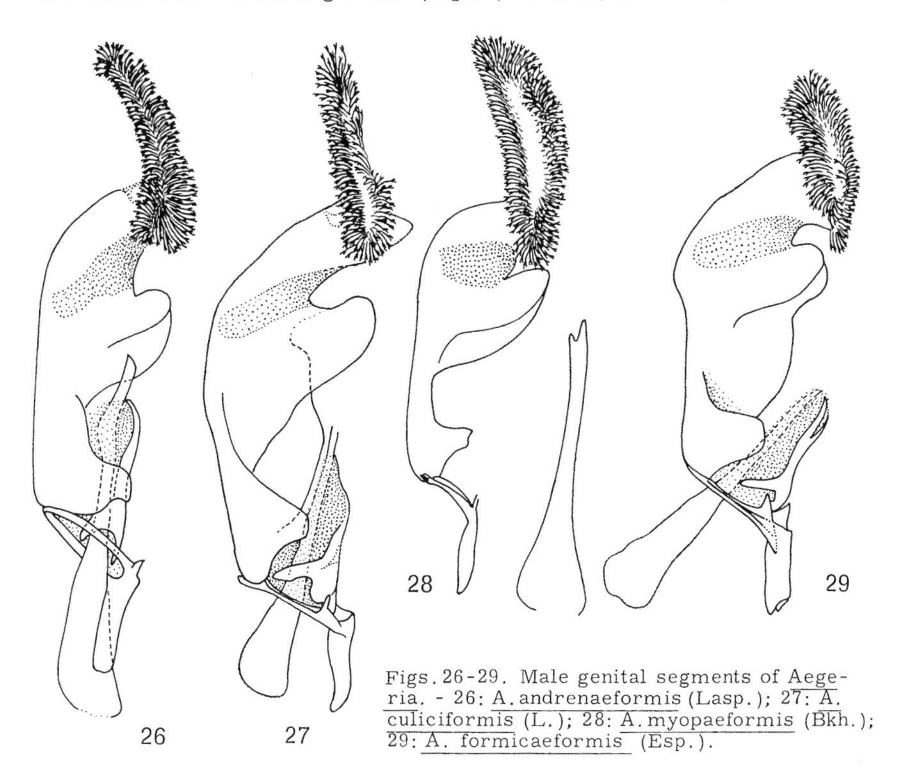

Figs. 26-29. Male genital segments of Aegeria. - 26: A. andrenaeformis (Lasp.); 27: A. culiciformis (L.); 28: A. myopaeformis (Bkh.); 29: A. formicaeformis (Esp.).

front of ostium fused into an anteriorly concave bridge; behind the ostium the plates converge markedly towards each other. Ductus bursae with very short, collar-like sclerotized antrum. - Pupal frontal process (Fig. 74) a low, curved ridge.

All Scandinavian countries, northwards beyond the Arctic circle. - Widely distributed in temperate W. Palaearctic region.

The adults occur in June and July. In early morning newly emerged specimens are found on the trunks of old Betula, the host tree. After copulation, which takes place later in the morning, both sexes visit flowers, exuding birch sap or artificial sweet baits. In sunny weather females will fly around laying eggs until about 15oo. The eggs are laid in old emergence holes or bark crevices in the lower part (0-2m) of old trunks. The larvae bore between bark and wood; the galleries are very irregular (Fig. 129) and may extend below ground level or onto the shady side of the tree. The larva hibernates twice. In April or May it bores a horizontal chamber in the bark and here constructs a tough cocoon (Fig. 130) mixed with small, rounded wood-scrapings; the future emergence hole is covered with a thin layer of bark. According to British observa-

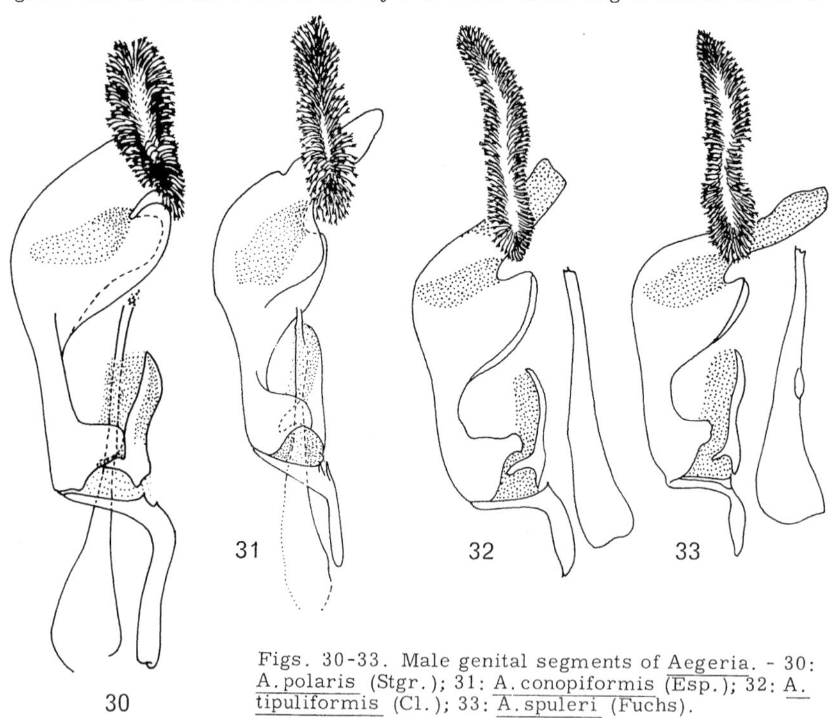

Figs. 30-33. Male genital segments of Aegeria. - 30: A. polaris (Stgr.); 31: A. conopiformis (Esp.); 32: A. tipuliformis (Cl.); 33: A. spuleri (Fuchs).

tions (Classey et al., 1946; Edelsten et al., 1961) the larva makes the cocoon in summer after the second hibernation, but does not pupate until next spring, i.e., after a third hibernation. We are not aware of similar records from the continent. The species is associated with old, sun-exposed birch trees, which may stand on moist as well as on dry soils.

Larvae or pupae may be obtained from trees where infestation is evidenced by the presence of old emergence holes (Fig. 128). Pupae may be found by brushing the trunk surface with a steel brush, thereby breaking the thin bark coverings above the cocoons; the latter can then be removed by carefully cutting out the surrounding piece of bark, about 6 cm in diameter. Alternatively the full

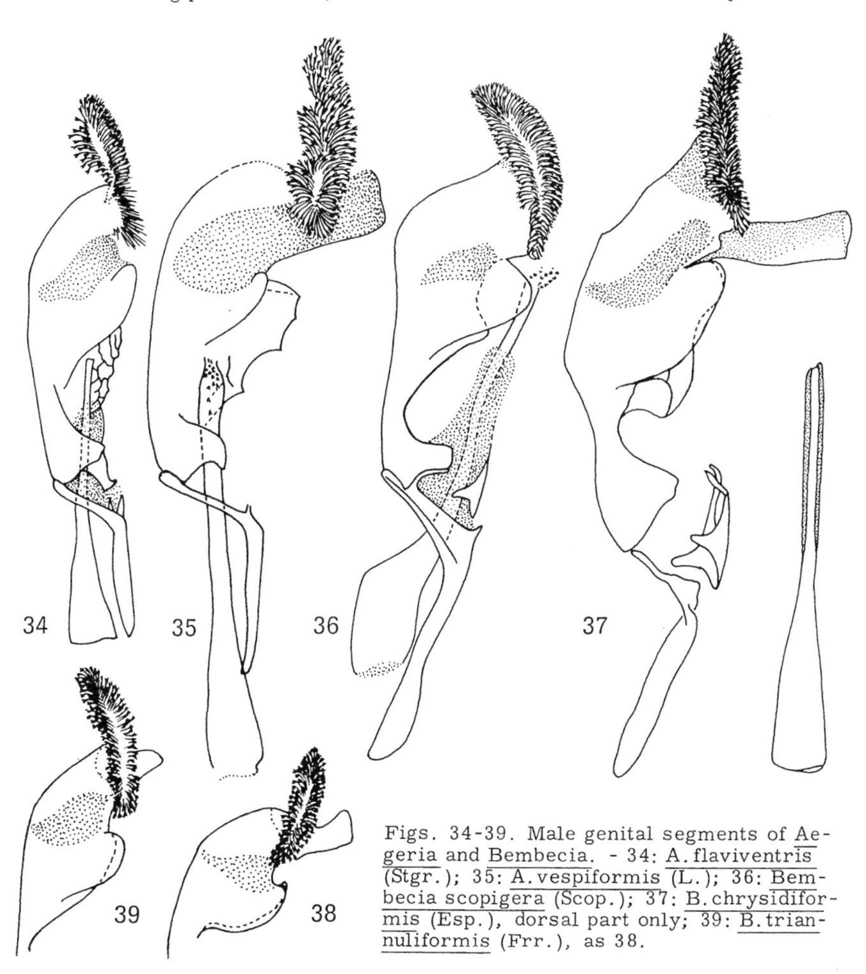

Figs. 34-39. Male genital segments of Aegeria and Bembecia. - 34: A. flaviventris (Stgr.); 35: A. vespiformis (L.); 36: Bembecia scopigera (Scop.); 37: B. chrysidiformis (Esp.), dorsal part only; 39: B. trianuliformis (Frr.), as 38.

grown larvae may be found in early spring by systematically removing bark where frass is extruded. The larvae thus obtained are more hardy than the pupae, but because of the irregularity of the galleries and the similar appearance of new and old frass, the larvae are difficult to locate and may easily be damaged. Collected larvae will accept sawdust for cocoon-making.

7. AEGERIA MESIAEFORMIS (Herrich-Schäffer, 1845)
Figs. 24, 41, 58, 75, 91.

Sesia mesiaeformis Herrich-Schäffer, 1845: 65, 74-75.

Facies generally similar to A. spheciformis; distinguishable by white laterofacial scales, presence of dorsal band on abdominal segment IV (in addition to that on II) and much deeper yellow colour of markings. Antennae in both sexes

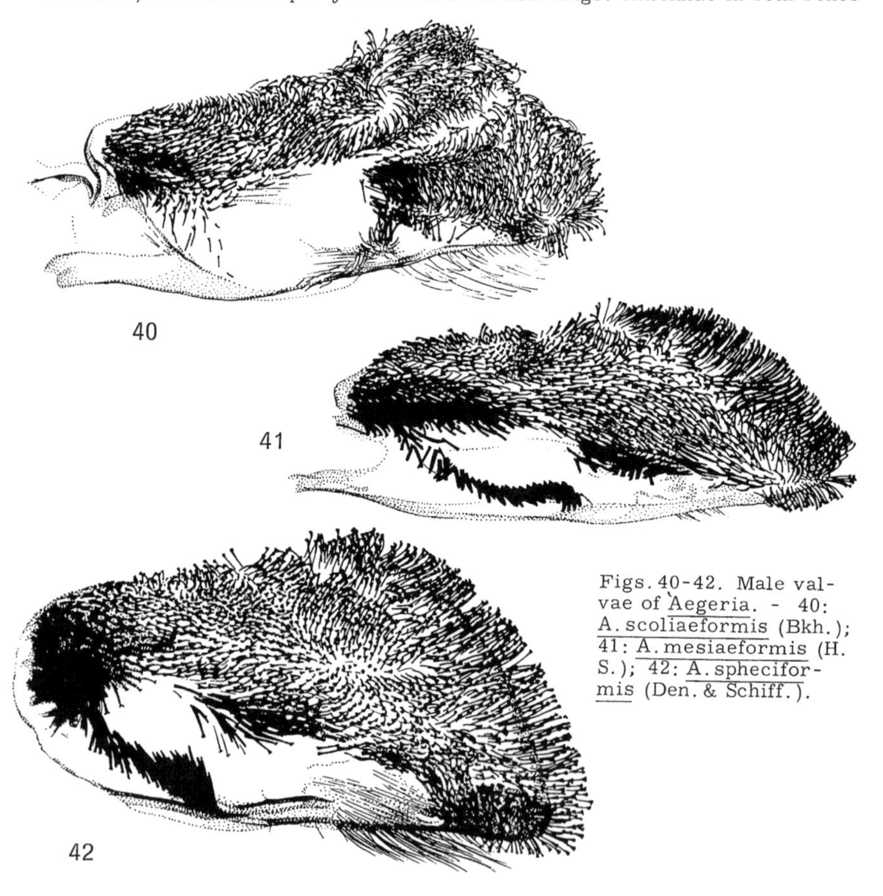

Figs. 40-42. Male valvae of Aegeria. - 40: A. scoliaeformis (Bkh.); 41: A. mesiaeformis (H. S.); 42: A. spheciformis (Den. & Schiff.).

with yellow subapical band. Tegulae edged yellow. Hind tibia bright yellow with black markings. FW 12-15 mm. - Male genitalia (Figs. 24, 41): crista sacculi ventral, basally with diagonal curvature, covered with narrow scales. Crista gnathi long and high. Phallus with prominent subapical process. - Female genitalia (Fig. 58): lateral plates of segment VIII anteroventrally fused as in *A. scoliaeformis*, but bridge longer, its anterior margin straight. Ductus bursae with funnel-shaped, sclerotized antrum. - Pupal frontal process (Fig. 75) prominent, reaching distinctly beyond vertex.

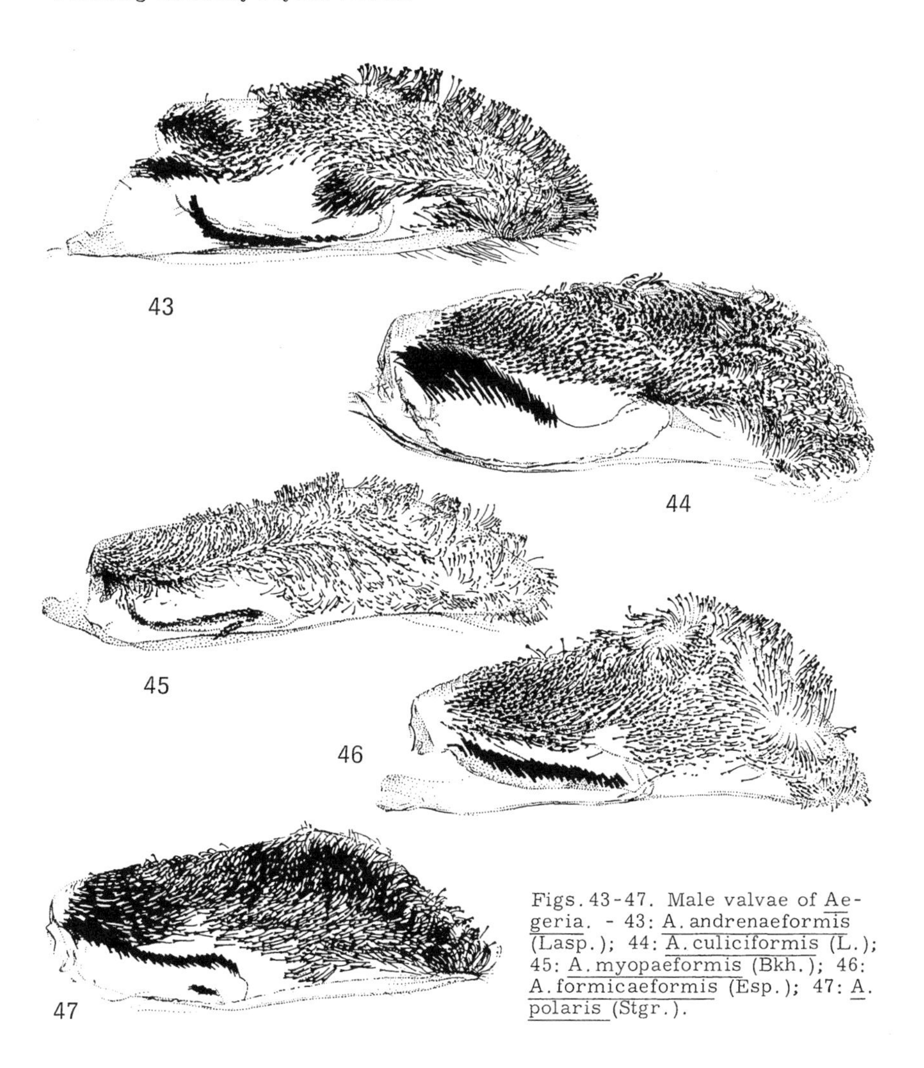

Figs. 43-47. Male valvae of *Aegeria*. - 43: *A. andrenaeformis* (Lasp.); 44: *A. culiciformis* (L.); 45: *A. myopaeformis* (Bkh.); 46: *A. formicaeformis* (Esp.); 47: *A. polaris* (Stgr.).

Southernmost provinces of Finland only. - Central Palaearctic region. European occurrences very disjunct, including besides the Finnish localities the Wolgograd area, Danube basin and Yugoslavia (Saramo, 1973).

The adults occur in June and July; in daytime they may be found resting on the trunks of the host trees, old Alnus glutinosa. The eggs are laid in bark crevices 0-2 m above groundlevel. The larva makes irregular tunnels between bark and wood. Its presence is revealed by the extruded frass which adheres to the bark by silken threads. The larva hibernates twice and pupates in May. Pupation takes place in a cocoon mixed with bark scrapings; the future emergence hole is covered by a thin layer of bark.

The habitats are groups of old Alnus glutinosa.

Since the bionomics of A. mesiaeformis closely resembles that of scoliaeformis, the former species may be collected by means of the techniques outlined for the latter species.

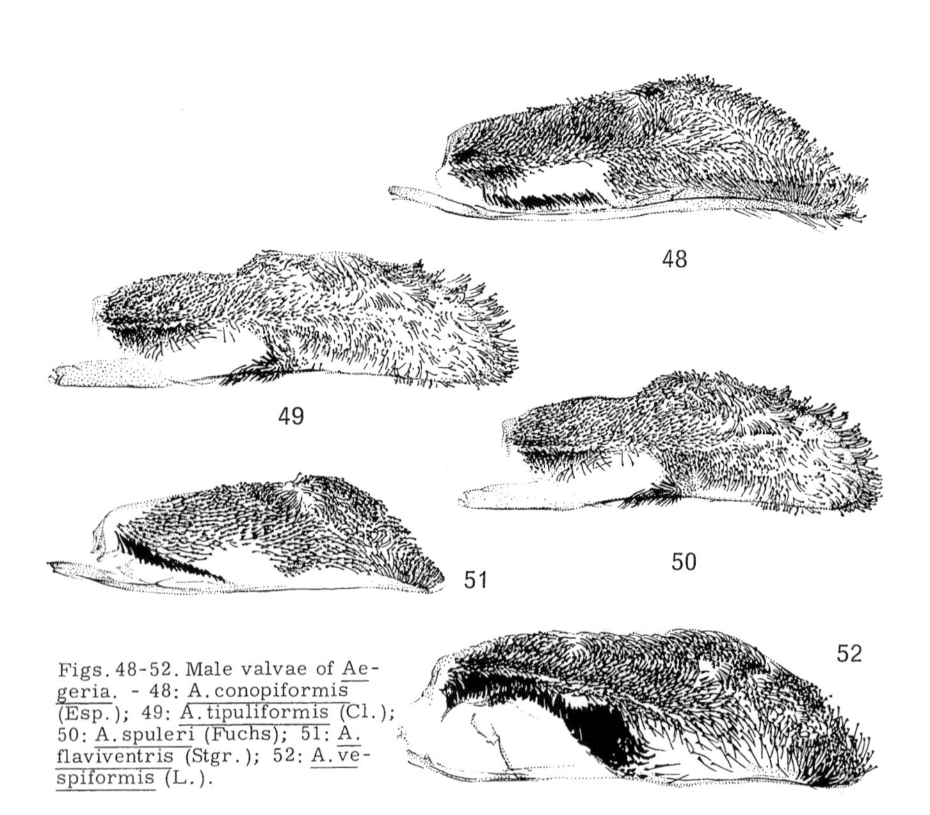

Figs. 48-52. Male valvae of Aegeria. - 48: A. conopiformis (Esp.); 49: A. tipuliformis (Cl.); 50: A. spuleri (Fuchs); 51: A. flaviventris (Stgr.); 52: A. vespiformis (L.).

8. AEGERIA SPHECIFORMIS (Denis & Schiffermüller, 1775)
Figs. 25, 42, 59, 76, 92, 131-132.

Sphinx spheciformis, Denis & Schiffermüller, 1775: 306.

Differs from the other large Aegeria species in having face entirely dark, dorsal band on abdominal segment II only and very pale yellowish-white colour of light markings. Antennae in both sexes with light subapical band. Tegulae edged light. Hind tibia dark, surroundings of proximal spurs light. FW 12-15 mm. - Male genitalia (Figs. 25, 42): crista sacculi diagonal with strong setae. Crista gnathi short, gnathos flaps unusual in being ventrally concave. Phallus externally very finely dentate in apical part. - Female genitalia (Fig. 59): highly distinctive, sclerotized and granulated lamella antevaginalis. Ductus bursae sclerotized beyond origin of ductus seminalis; antrum funnel-shaped.

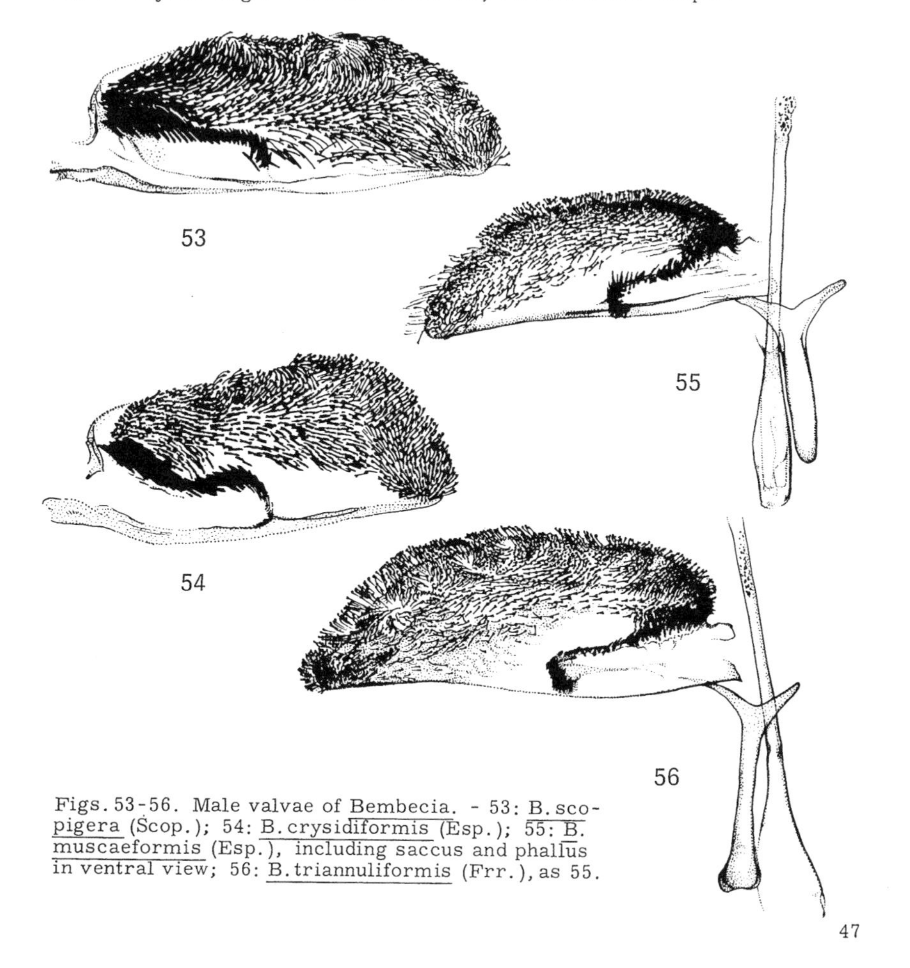

Figs. 53-56. Male valvae of Bembecia. - 53: B. scopigera (Scop.); 54: B. crysidiformis (Esp.); 55: B. muscaeformis (Esp.), including saccus and phallus in ventral view; 56: B. triannuliformis (Frr.), as 55.

Larval labrum with distal margin concave (straight or convex in all other Scandinavian Aegeria except polaris). Pupal frontal process (Fig. 76) very prominent, with tongue-shaped apex and raised lateral margins.

All Scandinavian countries, northwards beyond the Arctic circle. - Widely distributed in temperate Palaearctic region.

The adults occur in June and July and may in daytime be found on trunks, leaves, flowers or artificial sweet baits. The host trees are Alnus glutinosa, less typically A. incana or Betula. The eggs are laid in ground-near bark crevices. Suckers or young trees may be infested as well as older ones and even stumps. The young larvae bore between bark and wood; their tunnels are lined with silk and their brown frass is frequently found hanging down from the tree in a delicate web. In small trees (diameter 5 cm or less) the larva may make its tunnel all the way around the periphery in the border layer between bark and wood, or it may descend far down into the root. After the first hibernation it goes into the wood and bores upwards (Figs. 131-132); the frass now extruded is dark brown. After the second hibernation the future emergence hole is made, 0-50 cm above groundlevel; it is covered by a very thin layer of bark. Pupation takes place without construction of any cocoon. Preferred habitats are moist woods or swamps with ample growths of the food trees.

Trees infested by spheciformis are recognizable by the presence of the light second-year frass on the ground below. In spring the emergence holes may be exposed by brushing the lower part of the trunk with a steel brush, thereby breaking the thin bark covers. Young trees or suckers containing tunnels are liable to break if bent. Because of the mobility of the pupa the whole plant must be cut near groundlevel to ensure successful rearing. Less damage is made by obtaining larvae or pupae from stumps; often several pupae occur together in the same stump. Larvae and pupae may be placed in moist sawdust for rearing.

AEGERIA ANDRENAEFORMIS (Laspeyres, 1801)

Figs. 26, 43, 60, 93, 127.

Sesia andrenaeformis Laspeyres, 1801: 20.
Sphinx anthraciniformis Esper, 1798: 29 (see note below).

In general facies somewhat intermediate between the large Aegeria species (scoliaeformis, mesiaeformis, spheciformis) and the tipuliformis-group. Antennae of both sexes, face and tegulae uniformly dark. FW ETA relatively narrow, not including the proximal area between branches of radial fork. Abdomen dorsally with light bands on segments II and IV, anal tuft apically suffused with light to variable extent, light coloration varying from yellow to bright orange. Hind tibia dark with narrow yellow apex; light area surrounding proximal spurs very small, often absent. FW 8-11 mm. - Male genitalia (Figs. 26, 43): crista sacculi an indistinctly delimited line of folded cuticle with a proximally diagonally curved row of scales. Crista gnathi long and high, convex. Phallus subapically with dorsal curvature. - Female genitalia (Fig. 60): antevaginal area of venter VIII moderately sclerotized, transversely folded. Ductus bursae with distinctly sclerotized, funnel-shaped antrum with convex outline.

Pupal frontal process with transverse ridge.

Not yet recorded from Scandinavia. Scattered occurrences from S. England through C. Europe to Asia Minor and Central USSR.

The adults occur from the end of May to the beginning of July. The eggs are laid in bark crevices or in axils of Viburnum lantana and V. opulus; Sambucus ebulus is recorded as an alternative host. The larva bores centrally in the trunk or in a branch; it hibernates twice. Pupation takes place where the diameter of the stem permits the construction of a chamber at almost right angles to the surface (e.g., in a knot); no cocoon is formed. The future emergence hole is covered by a large round bark lid (Fig. 127), which is loosened along the periphery and which occasionally may fall off, leaving the pupa exposed (Rothschild, 1906). Preferred habitats are sunny wood-glades and hedges, at least in some C. European provinces only occurring on chalky soils (Bergmann, 1953).

Specimens may be located by searching for the somewhat hollowed bark lids covering the pupal chambers.

Note. An application is being made (Kristensen, in press, a) to the International Commission of Zoological Nomenclature that the hitherto used specific name andrenaeformis be protected against the forgotten anthraciniformis which has priority.

9-11. "Red-banded group".

Included under this heading are three medium-sized Aegeria-species which all have a broad, red band on abdominal segment IV. Moreover, they have white laterofacial scales, unicolorous dark tegulae and the males have high cristae sacculi on the valvae. The last-mentioned characters do not, however, pertain to all red-banded Aegeria outside Scandinavia.

9. AEGERIA CULICIFORMIS (Linné, 1758)
Figs. 27, 44, 61, 77, 94, 137.

Sphinx culiciformis, Linné, 1758: 493.

Usually the largest Scandinavian red-banded species (the specimen illustrated in Fig. 94 is, however, abnormally large). Like A. myopaeformis it has the FW apical area, anal tuft and hind tibia dorsally uniformly dark. Contrary to A. myopaeformis it has a distinct red suffusion on the FW base, red abdominal band ventrally complete and palps ventrally orange-red. FW 9-12 mm. - Male genitalia (Figs. 27, 44): crista sacculi very prominent, curved; in basal part with strong setae dorsally, in distal part naked. Crista gnathi long, angularly convex. Phallus simple, apically with minute cornuti. - Female genitalia (Fig. 61): venter VIII membranous. Ductus bursae with short, weakly sclerotized, sub-triangular antrum.

Pupal frontal process (Fig. 77) ventrally cone-like produced, with very small ridge apically.

One of the most common Scandinavian Sesiids, found in all countries and

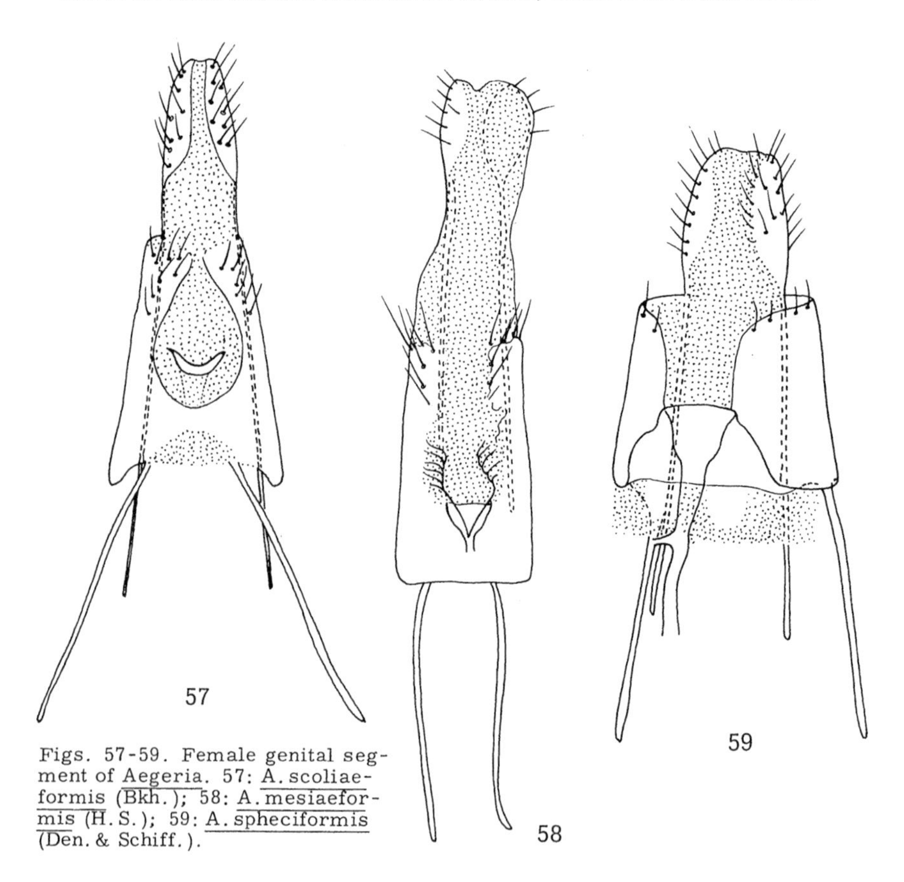

Figs. 57-59. Female genital segment of Aegeria. 57: A. scoliaeformis (Bkh.); 58: A. mesiaeformis (H. S.); 59: A. spheciformis (Den. & Schiff.).

ranging to northernmost provinces. - Widely distributed in the Palaearctic region and also found in N. America.

The adults occur in May and June. They visit flowers and may also be found resting on foliage or, particularly, stumps. The host is primarily Betula, but also not infrequently Alnus; in both it may occur together with Aegeria spheciformis. A. culiciformis is, however, the most polyphagous of our wood-boring Sesiids and may occasionally occur in a variety of quite unrelated trees, e.g., species of Quercus, Fagus, Ulmus, Prunus and Tilia. The eggs are laid in ground-near bark crevices in trunks or between bark and wood on the cut surface of stumps. The larva apparently hibernates twice. In the first year it makes tunnels between bark and wood whereas in the second it may enter the wood. The scrapings of this species are most characteristic long chips, which often protrude from the tunnels. In spring after the second hibernation the larva makes a tough cocoon mixed with the long chips (Fig. 137). In trunks the future emergence hole is covered by a silk or bark lid. Frequently many larvae occur together. Preferred habitats are sun-exposed trees and stumps in moors, heaths and woods, i.e., in moist as well as dry situations.

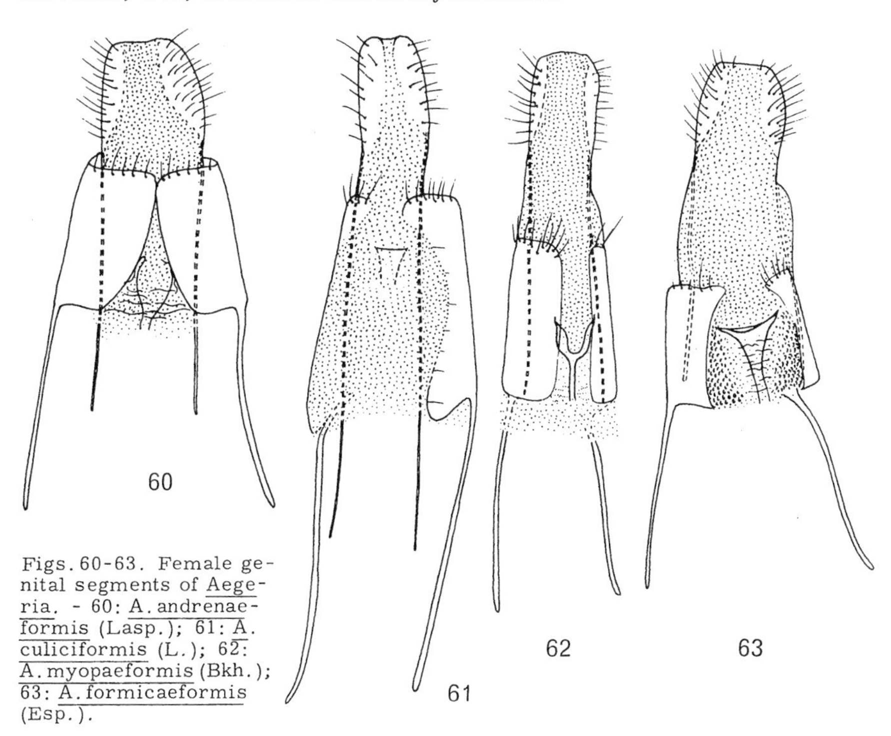

Figs. 60-63. Female genital segments of Aegeria. - 60: A. andrenaeformis (Lasp.); 61: A. culiciformis (L.); 62: A. myopaeformis (Bkh.); 63: A. formicaeformis (Esp.).

The presence of the species is most easily discovered by searching for the characteristic chips in stumps of Betula or Alnus; the chips may be found by removing the bark or they may occur in heaps on top of the stump. The pupa may be located immediately below the bark, but more often it is found in the wood where it may be removed by cutting out its surroundings.

10. AEGERIA MYOPAEFORMIS (Borkhausen, 1789)
Figs. 28, 45, 62, 95.

Sphinx myopaeformis Borkhausen, 1789: 169.

The species is very similar to A. culiciformis but smaller. Labial palps in female uniformly dark, in male with whitish ventro-median suffusion. FW base without red suffusion. Red abdominal band ventrally interrupted by dark ground-colour. FW 8-9 mm. - Male genitalia (Figs. 28, 45): crista sacculi curving diagonally at base, with scales dorsally and in a small subapical patch on the median side. Crista gnathi long, smoothly convex. Phallus with prominent subapical process. - Female genitalia (Fig. 62): venter VIII anteriorly very weakly sclerotized and transversely folded. Ductus bursae posteriorly moderately sclerotized and transversely folded. Ductus bursae posteriorly moderately sclerotized. Broad antrum with ventral wall very deeply and widely emarginate.

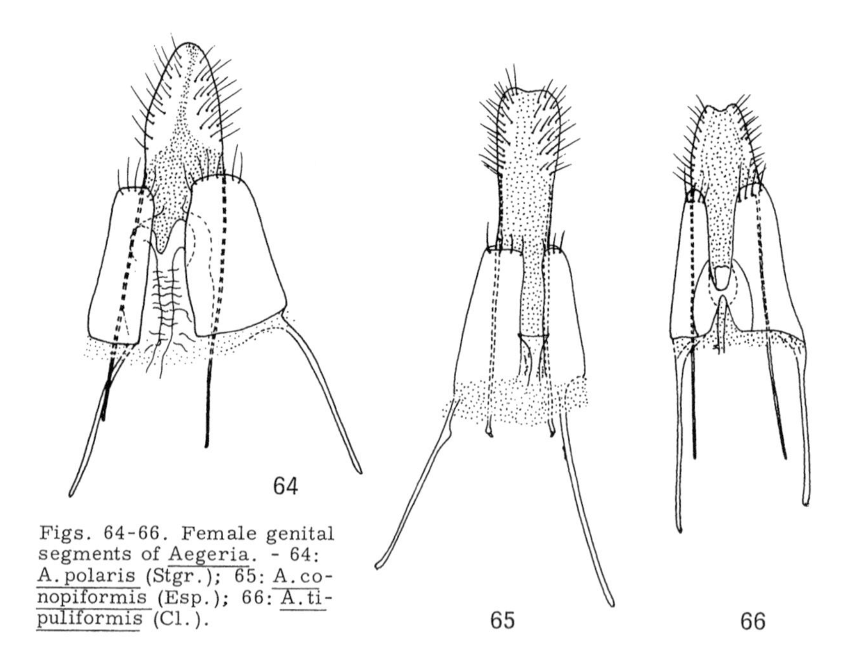

Figs. 64-66. Female genital segments of Aegeria. - 64: A. polaris (Stgr.); 65: A. conopiformis (Esp.); 66: A. tipuliformis (Cl.).

Pupal frontal process with transverse ridge, depressions on vertex very pronounced (Kemner, 1922).

In Scandinavia hitherto known only from South and Central Sweden (Scania, Uppland) and southernmost Norway. A single Danish record is unreliable (Kaaber, 1964). - From Great Britain through C. and S. Europe to Asia Minor and S.E. European USSR.

The adults occur in June and July. They visit flowers, but are more frequently observed on trunks of the host trees or in the vegetation below. The host is primarily Malus, but may also be species of Prunus, Cerasus, Crataegus and Sorbus; the entirely unrelated Hippophae has also been recorded as host. The eggs are laid in bark crevices, particularly in cankerous parts. The larvae bore between bark and wood; often many individuals are found close together. The frass is packed in the galleries, rarely extruded. The larvae hibernate twice. Pupation takes place in May or June in a loosely woven cocoon mixed with bark particles. The future emergence hole is covered with a thin layer of

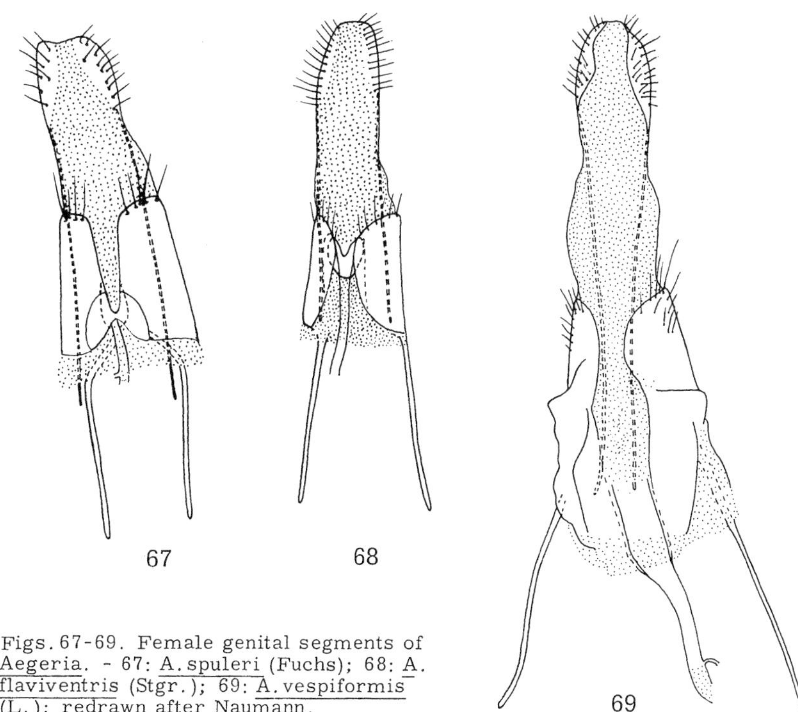

Figs. 67-69. Female genital segments of Aegeria. - 67: A. spuleri (Fuchs); 68: A. flaviventris (Stgr.); 69: A. vespiformis (L.); redrawn after Naumann.

bark. The pupal stage lasts about two weeks only. Preferred habitats are old orchards or growths of wild apples along roads or sunny wood glades.

Infested trees may be discovered by searching for old emergence holes (often with pupal cases) in cankerous swellings in the lower parts of the trunks. Inhabited tunnels may then be located by carefully removing the bark.

11. AEGERIA FORMICAEFORMIS (Esper, 1779)

Figs. 29, 46, 63, 78, 96, 133-136.

Sphinx formicaeformis Esper, 1779: 216.

Distinctive by bright reddish colour of FW apical area. Red abdominal band ventrally extending onto segment V and sometimes VI, anal tuft with white lateral suffusion. Newly emerged specimens with anterior abdominal segments dorsally suffused yellow, but the yellow scales are largely lost during flight. Hind tibia with whitish suffusion on inner side and in belts at mid-length and apex. FW 7-11 mm. - Male genitalia (Fig. 29, 46): crista sacculi diagonal, dorsally with narrow scales. Crista gnathi long, with ventral concavity, even pos-

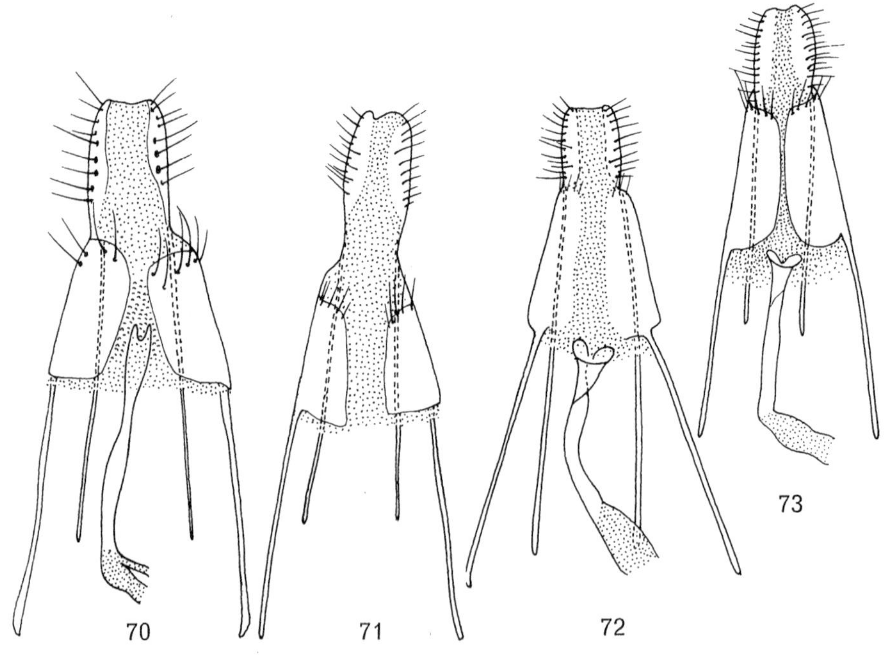

Figs. 70-73. Female genital segments of Bembecia. - 70: B. scopigera (Scop.); 71: B. chrysidiformis (Esp.); 72: B. muscaeformis (Esp.); 73: B. triannuliformis (Frr.).

teriorly higher than lateral flaps. Phallus simple, with minute apical cornuti. - Female genitalia (Fig. 63): venter VIII laterally coarsely granulated, medially more finely granulated and slightly folded transversely. Ductus bursae with sclerotized, sub-triangular antrum reaching origin of ductus seminalis; wall of antrum laterally produced in ostium.

Pupal frontal process (Fig. 78) very prominent, reaching distinctly beyond vertex.

All Scandinavian countries; in Sweden and Finland ranging northwards beyond Arctic circle, but in Norway known from southern provinces only. - Widely distributed in the Palaearctic region.

The adults occur in June and July; they may be found on foliage or flowers (e.g., of Rubus idaeus). Host plants are several species of Salix, and the effects of larval infestation vary somewhat according to the kind of host. The eggs are laid in bark crevices or in axils. The young larva bores between bark and wood. Having made a tunnel around the periphery of the branch the larva eventually bores into the wood. In spring, presumably after a single hibernation, it makes a pupal chamber beneath a thin layer of bark. The most frequently reported host appears to be Salix purpurea. On this tree as well as on S. alba, S. viminalis and S. triandra no deformation of infested branches appears. In S. repens and S. caprea the larva causes the formation of rather large, pyriform swellings (Fig. 133-134); in the first-mentioned plant the swellings are located fairly high (1-3 m), in the last-mentioned 0-30 cm above groundlevel (i.e., often hidden in the surrounding grass). The swellings are formed by frass being pressed out into the space between bark and wood (Fig. 133). The habitats are growths of the Salix species mentioned, i.e., on very different types of soil.

Immature stages are most easily found in those Salix species on which swellings are formed. On the other hosts the presence of larvae is rarely revealed by external sign, so branches of appropriate size (Fig. 135, diameter about 2 cm) must be selected by trial and error and carefully split. Infested branches are cut off and re-closed by means of rubber bands; each branch may contain several individuals.

12. AEGERIA POLARIS (Staudinger, 1877)

Figs. 30, 47, 64, 79, 97, 140.

Synanthedon polaris Staudinger, 1877: 175.
Sesia aurivilii Lampa, 1883: 127.

Distinctive by the extensive orange-red suffusion in FW non-transparent parts (except apical area), particularly along the hind margin. Abdomen dorsally with very narrow yellowish band on segment IV and similar, but still more indistinct bands on II and VI. In most specimens examined the yellow scales constituting the two last-mentioned bands have been lost or have become obsolete because of greasiness; only very fresh specimens have an appearence similar to the figured one. FW c. 10 mm. - Male genitalia (Figs. 30, 47): crista sacculi high, diagonal, dorsally with scales, subapically with medial patch of short stout setae. Crista gnathi short. Phallus simple, apically with small plate-like cornuti. - Female genitalia (Fig. 64): venter VIII very lightly sclerotized, finely granulated and conspicuously folded. Ductus bursae somewhat sclerotized for a considerable length; antrum only slightly widened, its latero-ventral wall continuous with a prominent rounded fold of venter VIII.

Pupal frontal process (Fig. 79) with low transverse ridge.

Arctic and subarctic regions of Norway, Sweden and Finland. - Not recorded outside N. Fennoscandia.

The life history has been fully described by Schantz (1959). The adults occur in July; they fly actively at noon on sunny days. The host plant is usually *Salix lapponum*, more rarely *S. caprea*. The eggs are laid in bark crevices or in axils and the larva initially mines in the bark. After the first hibernation it bores between bark and wood, extruding the frass to the exterior. After the second hibernation it goes into the wood and pith (Fig. 140); frequently it enters the root. In autumn the larva constructs a pupal chamber beneath a thin bark layer and located a few centimeters above groundlevel; no cocoon is made. Pupation does not take place until next June, i.e., the larva hibernates three times. The preferred habitats are south-facing slopes or bogs with *Salix lapponum*.

Infested plants may be discovered by removing grass and moss around the stems, thereby exposing the extruded frass. For rearing, the upper part of the root should be taken along with the lower part of the stem.

Note. Bartel (1912), who examined the type of *Sesia aurivilii*, claimed to have confirmed its specific validity, but a re-examination by Schantz (1959) conclusively showed it to be identical with *polaris*. The suggestions by previous authors that *Betula nana* is a host of the species have not been confirmed.

The *tipuliformis*-group

The species included under this heading (*conopiformis*, *tipuliformis*, *spuleri*, *flaviventris*) have a considerable resemblance to each other. They are all small and have narrow, but distinct yellow bands on abdominal segments II, IV, VI and (males, except flaviventris) VII. Antennae dorsally unicolorous dark, ventrally more or less suffused with yellow. Hind tibia dark, anteromedial part, apex and surroundings of proximal spurs yellow. FW apical area with lighter or darker brown suffusion between veins. Concerning genital structure the group is heterogenous.

AEGERIA CONOPIFORMIS (Esper, 1783)
Figs. 31, 48, 65, 98.

Sphinx conopiformis Esper, 1783: 213.

Distinguishable from other members of the group by yellow transverse band on metanotum, somewhat larger average size and usually brighter (more reddish than even the brightest tipuliformis) brown suffusion of the FW apical area. The height/width ratio of FW ETA is about 1, whereas in tipuliformis and spuleri this ratio is over 1, but some overlapping does occur. FW 8-10 mm. - Male genitalia (Fig.31, 48): crista sacculi ventral, curving slightly diagonally at base, with narrow scales. Crista gnathi short. Phallus subapically with small dorsal spines. - Female genitalia (Fig. 65): antevaginal part of venter VIII slightly sclerotized, longitudinally folded. Sclerotized, funnel-shaped antrum not quite reaching origin of ductus seminalis.

Not yet recorded from Scandinavia, but it may still remain to be discovered in the southern part of the area since its C. European distribution extends to the Baltic Sea. - From Belgium and France through C. and E. Europe to C. Asia.

The adults occur from June to August; they rest on foliage or on the lower part of the trunks of old Quercus, the principal host. Viscum album has been recorded as an alternative host. The eggs are laid in cankerous parts of the trunk or in stumps. The habitats of the larva are very similar to those of the other Quercus-feeding species, A. vespiformis (p. 62). It differs from the latter in making tunnels with a diameter distinctly larger than its own; moreover, it is stated to be more apt to extrude the frass from the galleries. The pupal chamber is spherical; the future emergence hole is covered by a thin bark lid. Pupation takes place in a delicate cocoon which contains frass particles. Preferred habitats are sun-exposed trunks or stumps of oak in open woods, parks, etc.
To reveal the emergence holes the bark should be bruched with a steel brush.

13. AEGERIA TIPULIFORMIS (Clerck, 1759)
Figs. 32, 49, 66, 80, 100-101, 141-142.

Sphinx tipuliformis Clerck, 1759: Pl. 9, 1.
Sphinx salmachus Linné, 1758: 493 (see note below).

Usually smaller than A. conopiformis. Metanotum unicolorous dark. FW ETA higher than wide, externally normally almost straight or slightly convex, very rarely concave; apical area with light suffusion more dull than in A. conopiformis. Postvertical and postocular scales as well as tegular inner margin yellow. FW 7-9 mm. - Male genitalia (Figs. 32, 49): crista saculi ventral, not distinctly elevated, the sensillae merging into those of sensory field. Crista gnathi relatively long, highest posteriorly. Phallus smooth, apically with minute processes. - Female genitalia (Fig. 66): lateral plates of segment VIII anteroventrally fused into a curved and sharp-edged antevaginal bridge, the surroundings of which are markedly sclerotized and pigmented. Antrum wide and membranous, but "roof" of the ostium with a distinct sclerotization. Ductus bursae sclerotized and straight between antrum and origin of ductus seminalis.

Pupal frontal process (Fig. 80) prominent and distinctly bifid.

One of the most familiar Sesiids in all Scandinavian countries; northwards

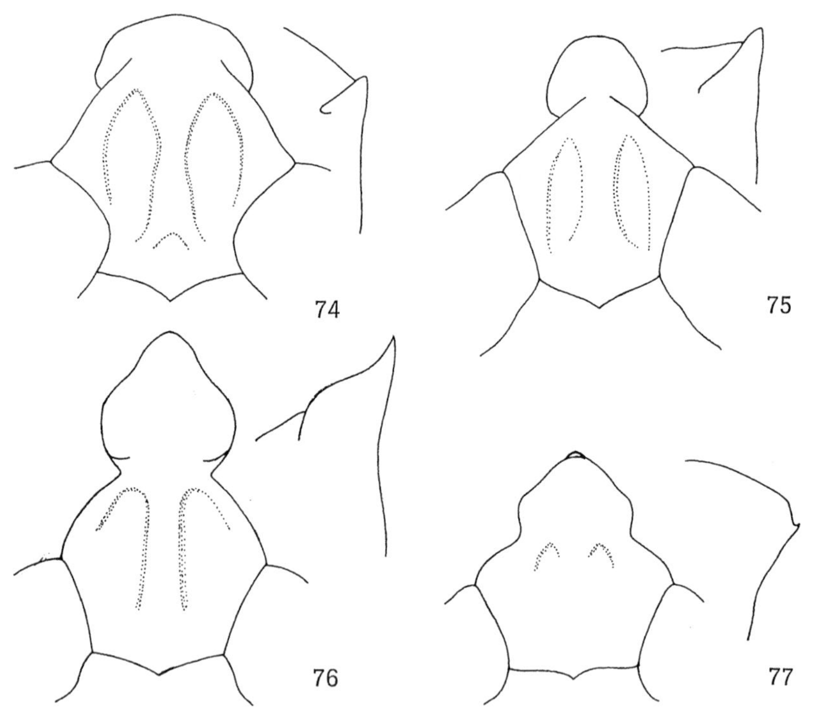

Figs. 74-77. Pupal frontal process and vertex of Aegeria; dorsal view and lateral profile. - 74: A. scoliaeformis (Bkh.); 75: A. mesiaeformis (H. S.); 76: A. spheciformis (Den. & Schiff.); 77: A. culiciformis (L.).

ranging beyond Arctic circle. - Widely distributed in the Palaearctic region and introduced into N. America, Australia and New Zealand.

The adults occur in June and July; they are often seen on foliage or flowers of the food plants. The moths fly actively in bright sunshine, activity stopping abruptly if the sun is covered by clouds. Food plants are species of Ribes, primarily R. nigrum, but also R. grossularia, R. rubrum, R. alpinum and R. uva-crispa. Rubus, Corylus and Euonymus have been recorded as occasional alternative hosts. The eggs are laid in axils or in pith of cut branches. The larva is presumed to be annual. The young larva bores into the pith, where hibernation takes place. In spring the larva makes its way into the bark, where it pupates. The pupa is surrounded by a loose web only; the future emergence hole is covered by a thin layer of bark. The young shoots emerging a few centimeters below the cut surfaces left by pruning are favorite pupation sites (Fig. 141). Preferred habitats are gardens and plantations where the species may become a troublesome pest. In nature it seems to prefer Ribes nigrum, often found in moist situations inside woods.

The presence of the larva is indicated by frass on the exterior of the branches; larvae may occur near groundlevel as well as high in the bushes. Whole infested branches should be cut off for rearing; many individuals may occur together in a single branch.

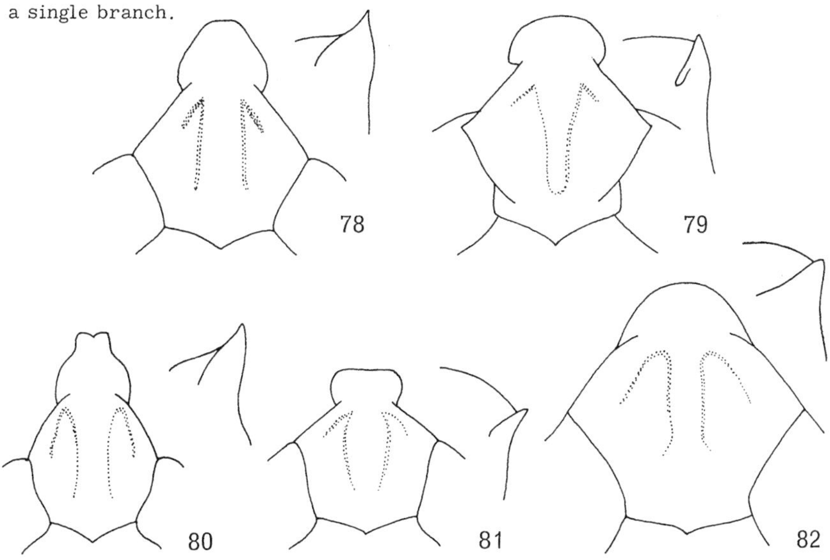

Figs. 78-82. Pupal frontal process and vertex of Aegeria; dorsal view and lateral profile. - 78: A. formicaeformis (Esp.); 79: A. polaris (Stgr.); 80: A. tipuliformis (Cl.); 81: A. flaviventris (Stgr.); 82: A. vespiformis (L.).

Note. An application is being made (Kristensen, in press, b) to the International Commission of Zoological Nomenclature that the hitherto used specific name tipuliformis should be protected against the forgotten salmachus which has priority.

AEGERIA SPULERI (Fuchs, 1908)
Figs. 33, 50, 67, 102.

Sesia spuleri Fuchs, 1908: 33.

A Juniperus-feeding species which is extremely closely related to A. tipuliformis and the taxonomic status of which is in much need of further study. In the standard handbooks (Hering, 1932; Forster & Wohlfahrt, 1960) A. spuleri is stated to differ from A. tipuliformis in being generally slightly larger and more robust, having FW ETA externally more distinctly convex, apical area less suffused light brown and anal tuft ventrally mixed with yellow hair-scales instead of being uniformly dark. However, these characters are all variable in both forms, and having compared a small material of C. European A. spuleri with a large number of A. tipuliformis, we must conclude that no reliable separation can be made on external characters. - Male genitalia (Figs. 33, 50): the single specimen examined has on the ventral surface of the phallus a low, rounded protuberance not observed in A. tipuliformis. - Female genitalia (Fig. 67): the few specimens examined are devoid of the sclerotized and pigmented areas in the "roof" of the ostium found in A. tipuliformis.

A. spuleri is by Forster & Wohlfahrt (1960) stated to be "In Mitteleuropa (which in their terminology includes Denmark) verbreitet, lokaler als die vorhergehende Art" (i.e. A. tipuliformis). The species has actually not been recorded from Denmark, but it should be sought for in S. Scandinavia. - Apparently not recorded outside C. Europe. It seems to be generally very rare and local, but it is probably largely overlooked.

The larva is stated to live in swellings on branches of Juniperus.

14. AEGERIA FLAVIVENTRIS (Staudinger, 1883)
Figs. 34, 51, 68, 81, 99, 138-139.

Sesia flaviventris Staudinger, 1883: 177.

Differs from other species in the group in having postvertical scales as well as tegulae uniformly black. FW apical area usually almost devoid of light suffusion; ETA markedly higher than broad and with concave external delimitation; discal spot in newly emerged specimens sometimes externally suffused dull orange, but the large scales are largely deciduous. Abdomen usually with a large yellow patch ventrally, reaching from segment IV to VI; occasionally this is dissolved into diffuse yellow markings along the hind margins of segments IV and V only. Male usually without band on VII. FW 6-9 mm. - Male genitalia (Figs. 34, 51): crista sacculi high, covered with scales. Crista gnathi broad, with reticulate pattern of low ridges. Phallus apically very slender and almost straight, with small cornuti. - Female genitalia (Fig. 68): venter VII membranous except for a small, strongly pigmented and V-shaped lamella antevaginalis bordering the deeply emarginate posteroventral margin of the antrum. "Roof" of ostium sclerotized. Ductus bursae with a long, strongly sclerotized and almost straight part posteriorly.

Pupal frontal process (Fig. 81) a low transverse ridge.

In Scandinavia known only from Denmark (Bornholm) and Finland (S. and S. E. provinces). - Scattered records from England and the N. European lowlands to USSR; also recorded from Roumania. The species is probably largely overlooked; the Danish and most of the Finnish populations have been discovered in the 1960's only.

The adults occur in July and may be found on foliage or branches of the food plants, Salix repens and S. aurita; Bartel (1912) supposed the species to be monophagous on S. caprea, but examinations of this plant in Denmark have so far yielded no result, and similar experiences were recorded from Pomerania by Urbahn (1939). The eggs are laid in axils on slender (diameter less than 1 cm) branches. The young larva makes a peripheral tunnel around the branch, and in this tunnel the first hibernation takes place. In the following spring the larva bores into the wood, pressing the frass into cavities between bark and wood whereby a somewhat pyriform swelling is formed on the branch (Figs. 138-139). The boring of the larva damages the conductive tissues, causing the leaves of the branch distal to the swelling to turn abdormally light green. The leaves produced in spring, after the second hibernation of the larva, exhibit a similarly distinctive coloration. Pupation takes place in June in a chamber immediately above the swelling, the pupa facing downwards. The future emergence hole is covered by a thin bark lid. Preferred habitats are growths of the food plants along the coast, in moorlands, along ditches, etc.

Infested branches may be found by searching for the abnormally coloured leaves or, in winter, the swellings.

15. AEGERIA VESPIFORMIS (Linné, 1761)
Figs. 35, 52, 69, 82, 103-104.

Sphinx vespiformis Linné, 1761: 289.

Distinctive by bright orange-red coloration of the external part of the FW discal and bright yellow ground colour of hind tibia; the latter has a black ventral streak and subapical band. Prominent (particularly in female) yellow dorsal bands on metanotum and abdominal segments II, IV (rarely V), and VI and (males only) VII. Anal tuft in male with medial and ventral yellow suffusion (often almost obsolete), in female yellow with distinct medial and ventral black suffusion. FW 7-11 mm. - Male genitalia (Figs. 35, 52): crista sacculi diagonal, its vestment of stout setae gradually merging into sensory field. Gnathos flaps very small, crista gnathi with peculiarly flattened and repeatedly bent ventral surface which is laterally demarcated by sharp edges. Phallus apically thickened and externally spiny, with row of small cornuti. - Female genitalia (Fig. 69): lateral plates of segment VIII anteroventrally continuous without formation of a discrete lamella antevaginalis. Ductus bursae ventrally sclerotized proximal to origin of ductus seminalis; antrum somewhat widened. Corpus bursae with diffuse signum.

Pupal frontal process (Fig. 82) with low transverse ridge.

In Denmark a single reliable record from S. Zealand. In Sweden a few records from Skåne, Öland and Uppland. Not in Norway and Finland. - Widely distributed in W. Palaearctic region.

The adults emerge during a long period, from May until August. They may be found resting on the lower parts of trunks or on stumps. The host tree is Quercus, less typically Fagus, Castanea and Tamarindus. The eggs are laid in cankerous swellings in lower parts of the trunks or in fresh stumps of old trees. The larva makes irregular tunnels between bark and wood, sometimes extruding the frass through bark crevices. It hibernates twice, both times enclosed in a loose web. The emergence hole, often located in a crevice, is covered with a thin bark lid. Preferred habitats are woodlands or parks with sun-exposed trunks or stumps of oak.

Larvae may be obtained by removing the bark of one- or two-year old stumps; the larvae will pupate in sawdust.

Note. Roovers (1964) found the genitalia of Dutch specimens to differ in details (arrangement of sensilla of crista sacculi in male, shape of ductus bursae in female) from those figured in Engelhardt (1946), and she assumed the latter to represent an "American form". However, A. vespiformis does not occur in the new world (it was figured by Engelhardt only because it is the type-species of the nominal genus Synanthedon) and the differences in question might easily have been produced as artifacts in preparation.

Genus *Bembecia* Hübner, 1819

Bembecia Hübner, 1819. Verz.bek.Schmett.: 128.
Type-species: Sphinx ichneumoniformis Denis & Schiffermüller, 1775= Sphinx scopigera Scopoli, 1763.
Dipsosphecia Püngeler, 1910, in Spuler: Schmett.Eur.2: 316.
Type-species: Sphinx ichneumoniformis Denis & Schiffermüller, 1775 (as above).
Pyropteron Newman, 1832. Ent.mon.Mag.1: 75.
Type-species: Sphinx chrysidiformis Esper, 1782.
Chamaesphecia auct., partim.

As discussed p.38 the genus Bembecia is here conceived in the same sense as in Bradley et al. (1972). Its members have the FW PTA short (usually not reaching the discal spot) or absent. Male genitalia as in Aegeria, with scopula androconalis and without crest delimiting sensory field of valve. Larvae stem- or root-borers in herbaceous plants.

Key to species of Bembecia

1	FW with orange suffusion, at least on discal spot and apical area	2
-	FW without orange suffusion	3
2 (1)	FW orange suffusion very extensive, in females often obliterating PTA and sometimes ETA. Abdomen with light bands on segments IV and VI (males sometimes VII) only (Fig.110)	chrysidiformis (Esp.)
-	FW orange suffusion less extensive, never concealing transparent areas. Abdomen usually with light bands on segments II-VI (-VII in males) (Figs.106-107)	16. scopigera (Scop.)
3 (1)	Associated with Armeria. Female PTA vanished (Fig. 109). Male saccus distally tapering (Fig.72), crista gnathi shorter than lateral flaps (Fig.38)	17. muscaeformis (Esp.)
-	Associated with Rumex acetosella. Female FW PTA distinct. Male saccus distally widened (Fig.73), crista gnathi longer than lateral flaps (Fig.39)	triannuliformis (Frr.)

Figs. 83-84. Pennisetia hylaeiformis (Lasp.)

FW PTA entirely basal to ATA.
ETA transversed by two veins only.
83: female, antennae simple; 84: male, antennae bipectinate.

Figs. 85-86. Sesia melanocephala Dalm.

Yellow abdominal bands relatively narrow.
85: male, antennae strongly pectinate; 86: female, antennae simple.

Fig. 87. Sesia apiformis (Cl.)

Patagial collar dark.
Tegulae anteriorly with prominent yellow patch.

Fig. 88. Paranthrene tabaniformis (Rott.)

FW largely covered with pigmented scales.

Fig. 89. Sesia bembeciformis (Hübn.).

Patagial collar yellow.
Tegulae unicolorous dark.

All figures in twice natural size

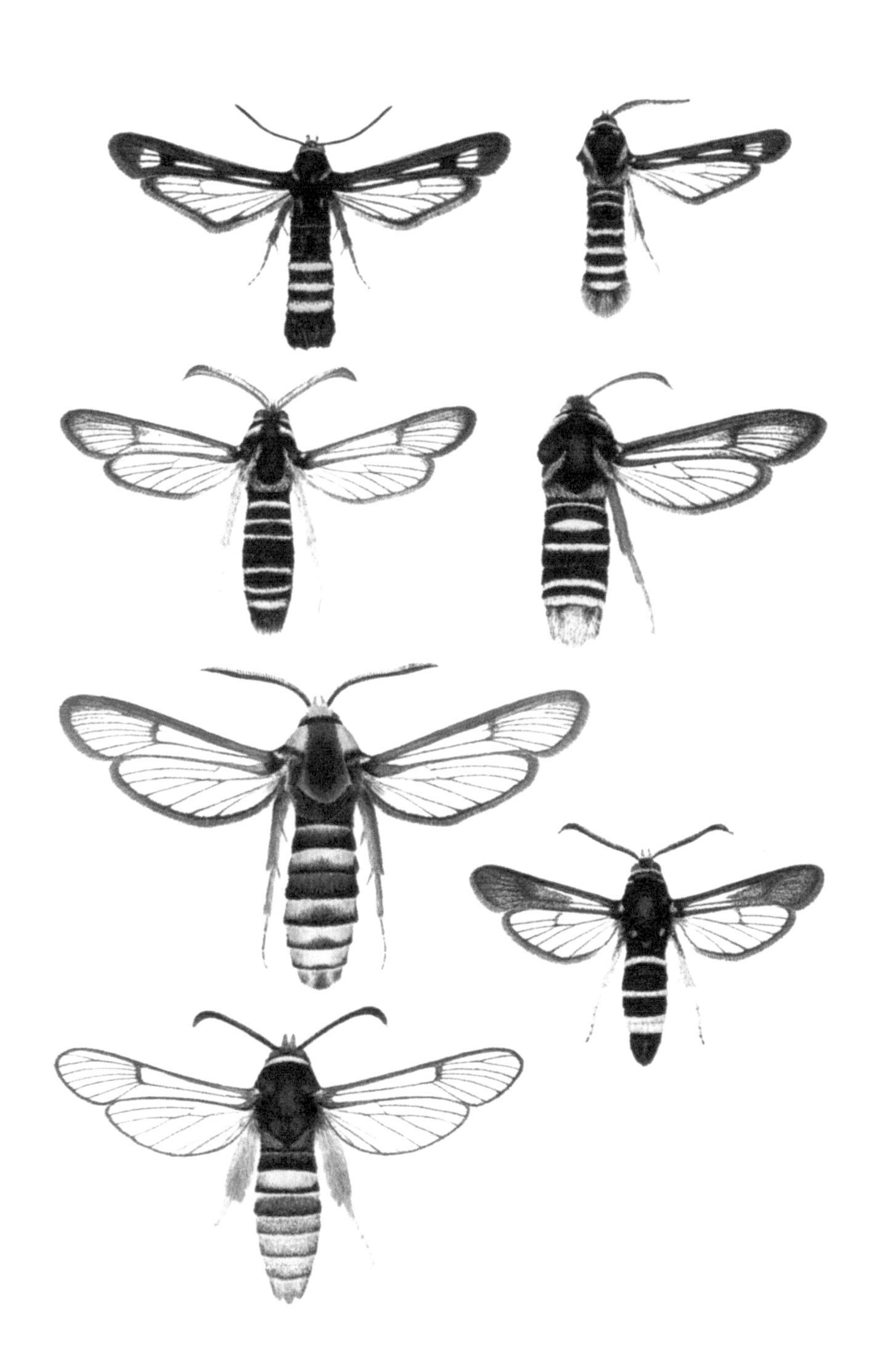

Fig. 90. Aegeria scoliaeformis (Bkh.)

Anal tuft orange.
FW discal spot very broad.

Fig. 91. Aegeria mesiaeformis (H. S.)

Hind tibiae bright yellow with black markings.
Yellow bands on abdominal segments II and IV.

Fig. 92. Aegeria spheciformis (Den. & Schiff.)

Hind tibiae predominantly dark.
Pale yellow band on segment II only.

Fig. 93. Aegeria andrenaeformis (Lasp.)

Antennae and tegulae unicolorous dark.
FW ETA relatively narrow.
Yellow bands on abdominal segment II and IV.

Fig. 94. Aegeria culiciformis (L.)

FW base suffused red.

Fig. 95. Aegeria myopaeformis (Bkh.)

FW base not suffused red.

Fig. 96. Aegeria formicaeformis (Esp.)

FW apical area suffused red.
Anal tuft laterally whitish.

Fig. 97. Aegeria polaris (Stgr.)

FW extensively suffused orange-red around transparent areas.
Abdominal bands often just barely discernible.

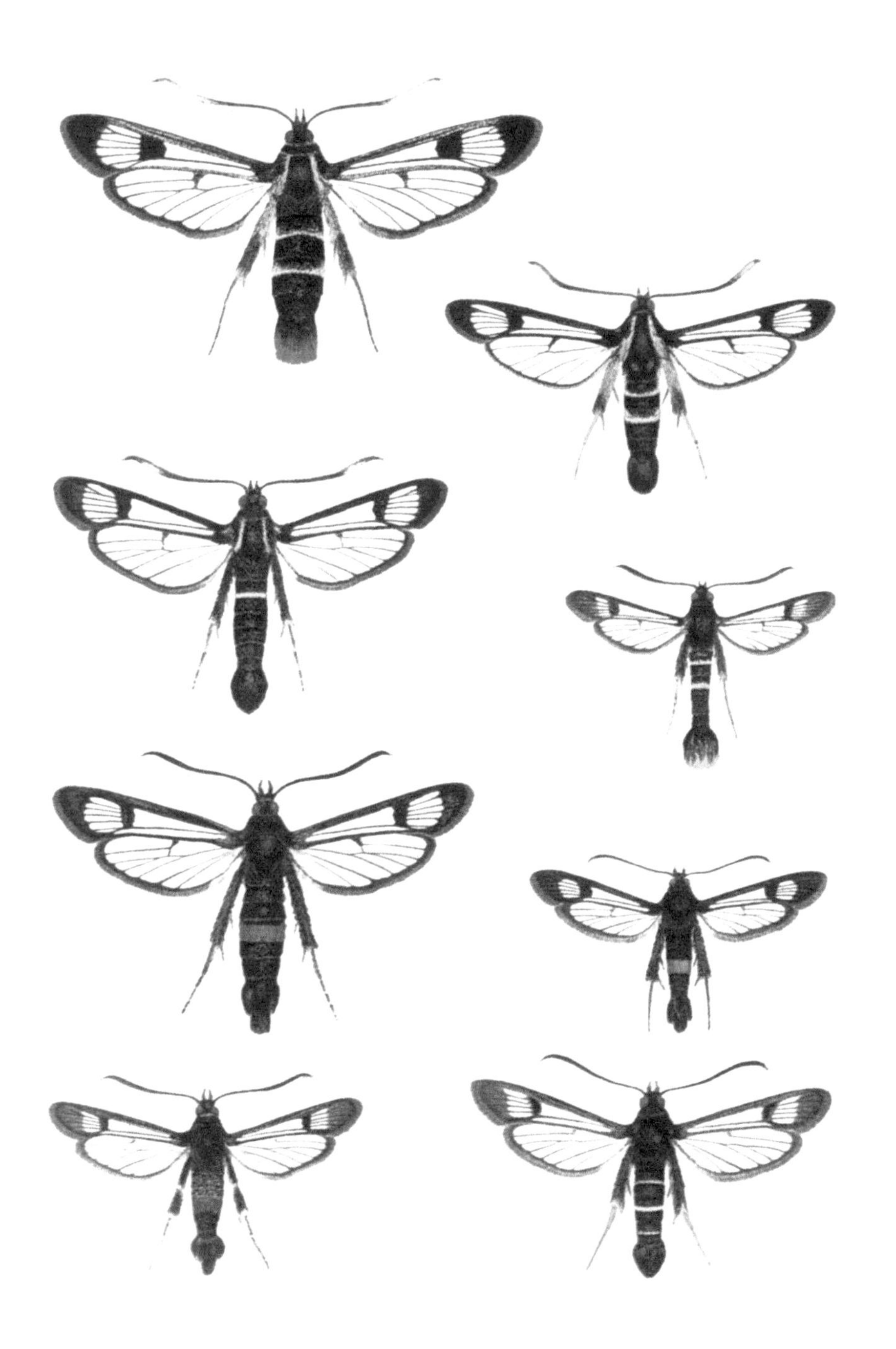

Fig. 98. Aegeria conopiformis (Esp.)

Metanotum with yellow band.

Fig. 99. Aegeria flaviventris (Stgr.)

Postvertical scales and tegulae unicolorous dark.
FW ETA externally concave.

Figs. 100-101. Aegeria tipuliformis (Cl.)

Postvertical scales and inner tegular margins yellow.
FW ETA externally not usually concave.

100: Male, four abdominal bands; 101: Female, three abdominal bands.

Fig. 102. Aegeria spuleri (Fuchs)

Externally not with certainty separable from A. tipuliformis.

Figs. 103-104. Aegeria vespiformis (L.)

Hind tibiae bright yellow with black markings.
FW discal spot largely bright orange-red.
103: Female, three abdominal bands, anal tuft largely yellow;
104: Male, four abdominal bands, anal tuft largely black.

Fig. 105. Bembecia triannuliformis (Frr.)

Male (figured) externally not with certainty separable from male B. muscaeformis.
Female PTA distinct.

Figs. 106-107. Bembecia scopigera (Scop.)

FW discal spot and apical area suffused orange.
Abdomen dorsally with yellow bands on III and (usually) V in addition to those on II, IV and VI (VII).
106: Female, no band on VII; 107: Male, with band on VII.

Figs. 108-109. Bembecia muscaeformis (Esp.)

108: Male externally not with certainty separable from male B. triannuliformis;
109: Female, FW PTA obsolete, covered with pigmented scales (in figured specimen grey).

Fig. 110. Bembecia chrysidiformis (Esp.)

FW non-transparent areas as well as hind tibiae bright orange.

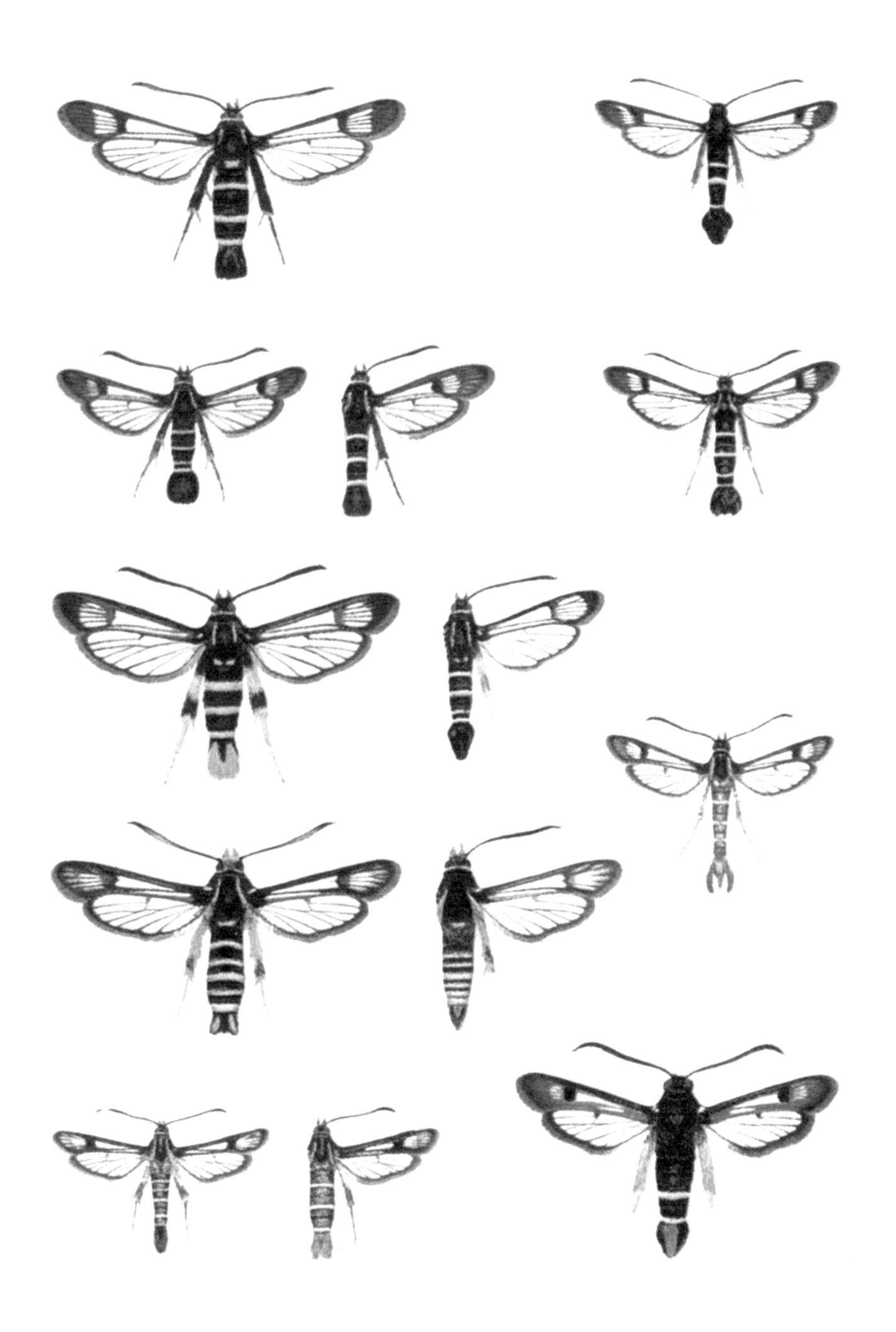

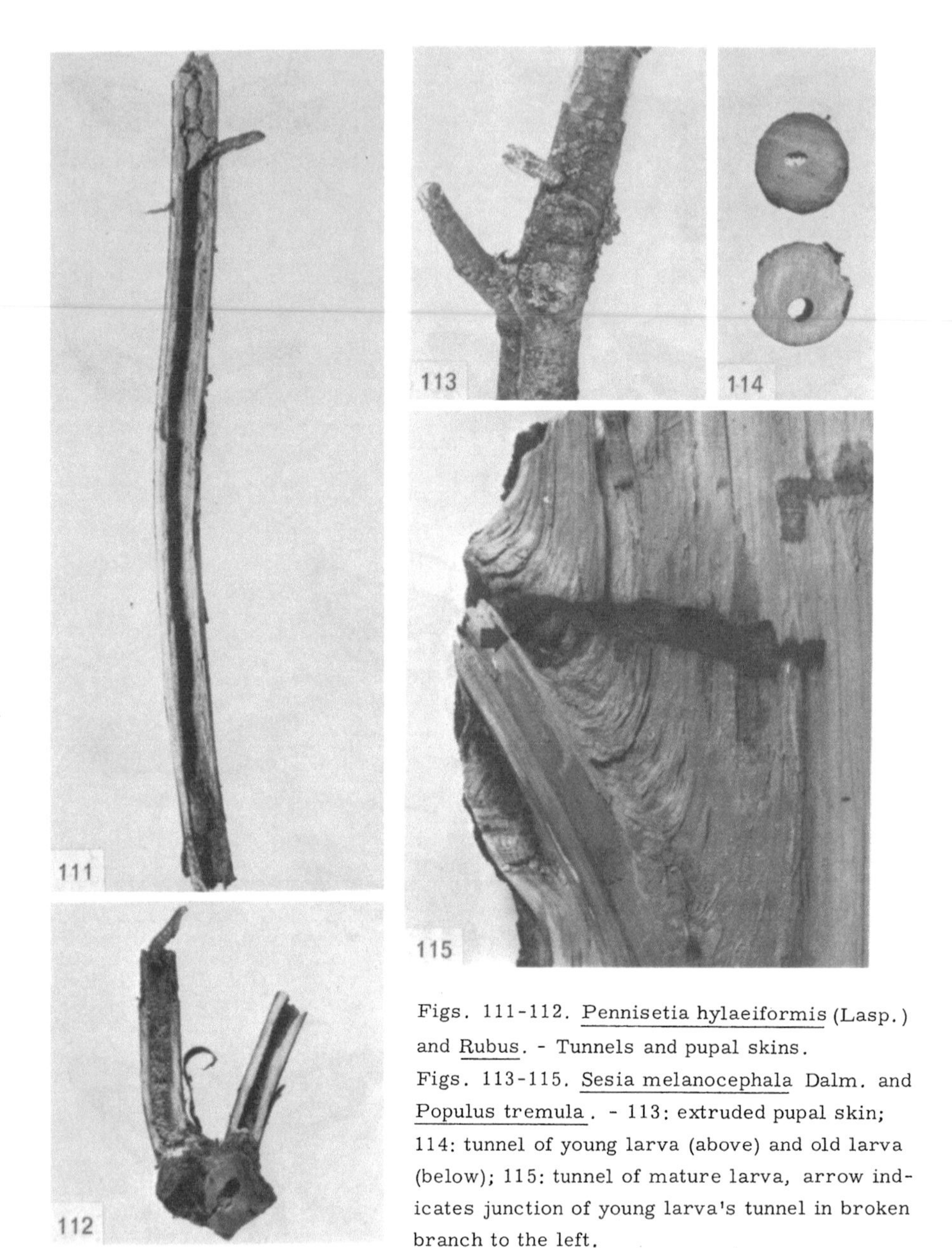

Figs. 111-112. Pennisetia hylaeiformis (Lasp.) and Rubus. - Tunnels and pupal skins.
Figs. 113-115. Sesia melanocephala Dalm. and Populus tremula. - 113: extruded pupal skin; 114: tunnel of young larva (above) and old larva (below); 115: tunnel of mature larva, arrow indicates junction of young larva's tunnel in broken branch to the left.

Figs. 116-120. Sesia apiformis (Cl.) and Populus. - 116: emergence hole (arrow) in lower part of trunk; 117: tunnels exposed, arrow indicates ground level; 118: covered emergence hole in bark; 119: same, lid removed; 120: cocoon on inner side of bark.

Figs. 121-123. Sesia bembeciformis (Hübn.) and Salix caprea. - 121: emergence hole (arrow); 122: terminal part of tunnel; arrow indicates position of emergence hole; 123: mature larva in cocoon.

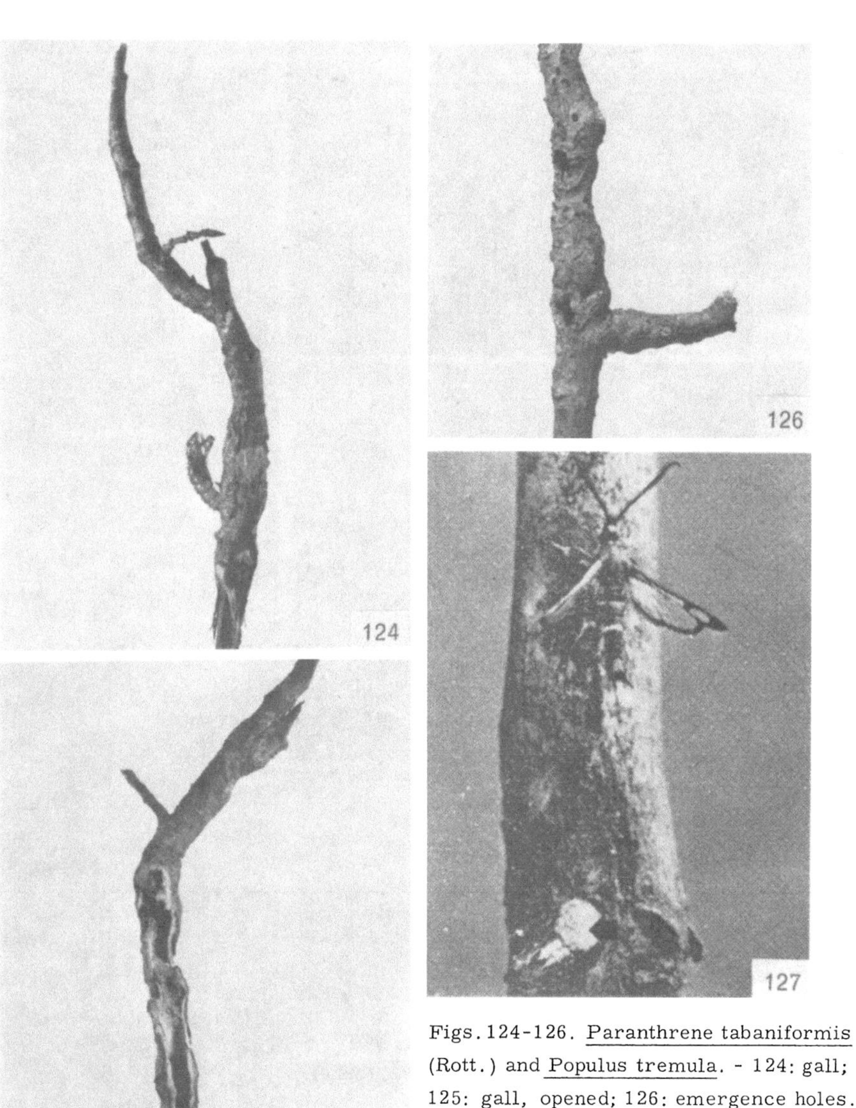

Figs. 124-126. *Paranthrene tabaniformis* (Rott.) and *Populus tremula*. - 124: gall; 125: gall, opened; 126: emergence holes. Fig. 127. *Aegeria andrenaeformis* (Lasp.) and *Viburnum*. - Newly emerged adult with extruded pupal skin and emergence hole lid (arrow); (after Classey et al., 1946).

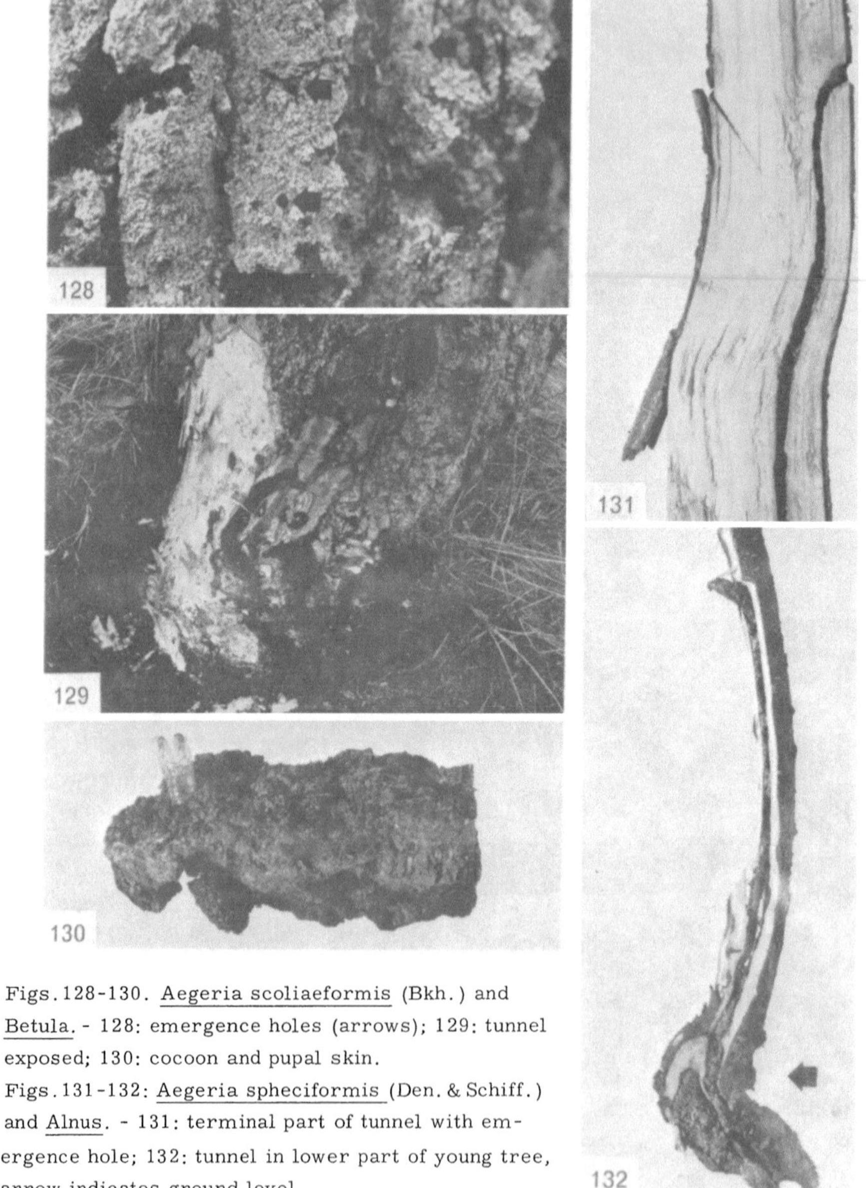

Figs.128-130. Aegeria scoliaeformis (Bkh.) and Betula. - 128: emergence holes (arrows); 129: tunnel exposed; 130: cocoon and pupal skin.
Figs.131-132: Aegeria spheciformis (Den. & Schiff.) and Alnus. - 131: terminal part of tunnel with emergence hole; 132: tunnel in lower part of young tree, arrow indicates ground level.

Figs. 133-136. Aegeria formicaeformis (Esp.) and Salix. - 133: opened gall in Salix repens; 134: gall and extruded pupal skin, Salix caprea; 135: tunnel of old larva, arrow indicates future emergence hole; 136: as 135, larva enlarged.

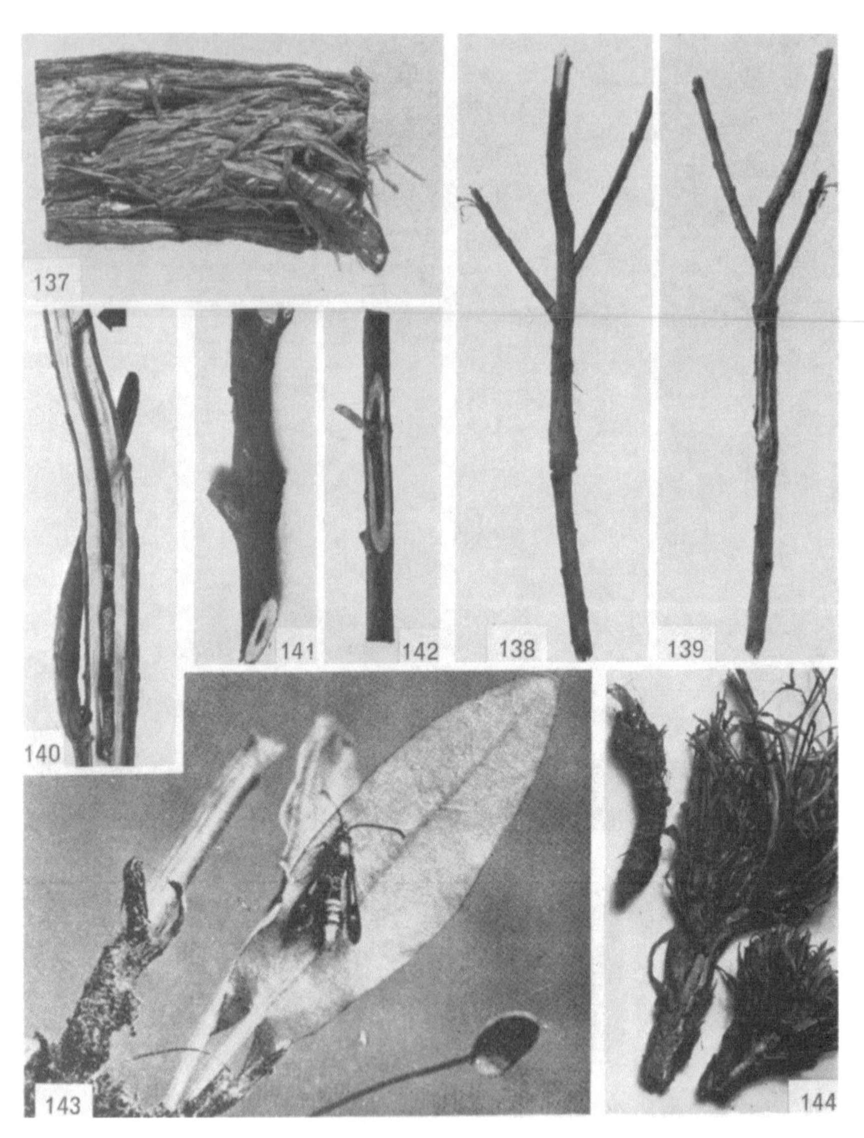

Fig. 137. Aegeria culiciformis (L.): cocoon and pupal skin. - Figs. 138-139. A. flaviventris (Stgr.) and Salix repens. - 138: infested twig; 139: as 138, opened. - Fig. 140. A. polaris (Stgr.) and Salix lapponum: tunnel with emergence hole (arrow). - Figs. 141-142. A. tipuliformis (Cl.) and Ribes.- 141: pupal skin extruded at base of cut shoot; 142: tunnel with pupal skin. - Fig. 143: Bembecia chrysidiformis (Esp.) and Rumex: newly emerged adult with pupal tube and skin (after Classey et al., 1946). - Fig. 144. B. muscaeformis (Esp.) and Armeria: pupal web (left) and larval tunnels (right).

16. BEMBECIA SCOPIGERA (Scopoli, 1763)
Figs. 36, 53, 70, 106-107.

Sphinx scopigera Scopoli, 1763: 188 n. 477.

Sphinx ichneumoniformis Denis & Schiffermüller, 1775: 44 n. 7.

Usually immediately distinguishable by the presence of a yellow dorsal band on each of the abdominal segments II-VI (-VII in males); in some specimens that of V is much reduced or entirely lacking. FW discal spot as well as internal part of apical area between veins suffused orange; orange suffusion in other parts of FW variable, but rarely prominent and never concealing the transparent areas. Proboscis short and lightly sclerotized. Second palp segment with prominent ventral tuft of long scales. Antennae dorsally unicolorous dark in male, with orange subapical band in female. Tegulae internally edged yellow and externally with white patch. Metanotum with yellow transverse band. Hind tibia orange-yellow, with black basal external stripe and subapical band. Anal tuft with yellow suffusion of variable extent. In some specimens the FW PTA may reach the discal spot; such specimens somewhat resemble Aegeria vespiformis but may be distinguished by proboscis structure, palp vestment, tegular patch and pattern of abdominal bands. FW 7-10 mm. - Male genitalia (Fig. 36, 53): crista sacculi ventral, very high, dorsally with stout setae of which the most distal are apically flattened. Crista gnathi long, angularly produced in posterior part. Phallus with smooth dorsal curvature and small cornuti. Saccus very long. - Female genitalia (Fig. 70): venter VIII membranous, finely granulated. Ductus bursae sclerotized in posterior part; antrum very gradually widened, at ostium with deep emargination in ventral wall.

Scattered records from Denmark and southern provinces of Fennoscandian countries, not ranging beyond 61° N. - Widely distributed in W. and C. Palaearctic region.

The adults occur from June to August; they may be found resting on the host plants which usually are Papilionaceae, particularly species of Anthyllis, Ononis and Lotus, more rarely Lathyrus, Medicago and Melilotus. The species has also been recorded to feed on Centaurea. The larva bores in the root, usually the primary root. The frass is not extruded from the gallery. Presumably the larva hibernates only once. In May or June it makes a long silken tube reaching from the root at a depth of 5-7 cm to the surface; the upper end of the tube is closed by a woven lid. Pupation takes place in the tube where the pupa can move freely. Preferred habitats are growths of the food plants in fields, gravel pits, etc., on poor or chalky soils.

The adult moths may be collected by sweeping the food plants. The immature stages are difficult to locate since infested plants show no definite symptoms. Whole plants may be dug up and the root carefully opened and searched for frass. Alternatively the root may be cut at some depth, since the larva will often be located in the upper part, particularly in sunny weather. Infested roots should be planted in a cage and frequently sprinkled; rearing is, however, often unsuccessful.

BEMBECIA CHRYSIDIFORMIS (Esper, 1782)
Figs. 37, 54, 71, 110, 143.

Sphinx chrysidiformis Esper, 1782: 210.

Immediately distinguishable by the extensive orange coloration of the FW (in females concealing PTA and sometimes ETA) and by the furry orange scale-covering of the tibiae. Proboscis well developed and sclerotized. Abdomen with greenish sheen and whitish-yellow dorsal bands on IV, VI and (males only, often almost invisible) VII. Anal tuft dorsally orange. - FW 7-10 mm. - Male genitalia (Figs. 37, 54): crista sacculi high, diagonal, curved; dorsally with very long and strong setae. Crista gnathi anteriorly V-shaped, each "arm" of the V broad, with sharp edges. Phallus slender, almost straight, with small cornuti. Saccus very long. - Female genitalia (Fig. 71): venter VIII and ductus bursae entirely membranous.

Not recorded from Scandinavia. - Distributed from S. England and Belgium through C. and S. Europe.

The adults occur in June and July. In sunny weather the moth visits flowers (various Compositae, Calendula, Rubus, etc.). The food plants are primarily various species of Rumex (notably R. acetosa and R. crispus) but the species has also been recorded to feed on various Compositae (species of Artemisia and Gnaphalium). The eggs are laid near the root and the larva bores into the root stock, where hibernation takes place. In May it may ascend the hollow stem and pupate here without formation of any cocoon; the future emergence hole remains covered by the stem epidermis. Alternatively pupation may take place inside a tough silken tube, which may be 3-5 cm long, is mixed with root scrapings and which stretches from the root onto the surface (Fig. 143). Infested plants are usually drying out and incapable of producing flowers. Preferred habitats are dry localities with ample growth of the food plants: fields, wood glades, chalky hills, slopes, etc.

In spring roots of plants showing the above-mentioned signs of infestation may be cut, about 5 cm below groundlevel, and searched for tunnels and frass. Plants containing larvae may be planted in a cage and frequently sprinkled.

17. BEMBECIA MUSCAEFORMIS (Esper, 1783)

Figs. 38, 55, 72, 108-109, 144.

Sphinx muscaeformis Esper, 1783: 217.

This species, and the following, are characterized by small average size as well as by the abdominal dorsal pattern: the transverse bands on II (not always present), IV, and VI are white, not yellow. Moreover, a more or less extensive yellow and/or white suffusion is present on all segments, sometimes forming a row of spots in the midline. Occasionally very faint bands on III and V are indicated by yellowish scales.

B. muscaeformis and triannuliformis are extremely closely related and difficult to separate. B. muscaeformis is on the average the smallest and has the abdominal light suffusion less distinctly defined. Female muscaeformis differ from both sexes of triannuliformis in not having any FW PTA, the entire postcubital area being covered with pigmented scales (which in the figured specimen are partly grey). The males of the two species are not generally separable on external characters. The difference between the two species concerning the the number of veins transversing the FW ETA (2-3 in muscaeformis, 4 in triannuliformis), repeatedly mentioned in European handbooks (latest Forster & Wohlfahrt,1960), is quite unreliable, since muscaeformis very commonly has the field transversed by 4 veins. Similarly it has not been possible to confirm the contention by Bartel (1912) that the antennae of male muscaeformis, contrary to those of triannuliformis, are equipped with short projections.

FW 5-8 mm. - Male genitalia (Figs. 38, 55): crista sacculi ventral, high, with stout hair-scales; distal end of crista somewhat less thickened than in triannuliformis. Crista gnathi reaching less far anterior than in triannuliformis. Saccus slightly tapering towards apex (in dorsal/ventral view). - Female genitalia (Fig. 72): venter VIII membranous. Ductus bursae sclerotized and with oblique folds of the wall. Antrum somewhat widened, posterodorsally produced into a pair of rounded lobes. The sclerotized portion of the ductus bursae is distinctly curved anteriorly (towards the right side of the animal).

In Denmark recorded from most provinces. In Sweden only known from Scania. Not in Norway and Finland. - From Great Britain through C. and E. Europe to southern European USSR.

The adults occur in June and July; they rest in daytime on the food plants, Armeria maritima (there are old records of Calluna as alternative host); in the evening they visit other flowers to feed. The eggs are laid near the roots of the food plant and the larva bores in the primary root (Fig. 144). Pupation takes place in the outer part of the root at surface level. The pupa is enclosed in a loose web containing frass particles (Fig. 144). The habitats are growths of Armeria maritima, i.e., in S. Scandinavia usually near the coast.

Adult moths may be swept from the food plants. Some authors state that plants infested by larvae of this species are recognizable by having fewer and more rapidly withering flowers. It must be noted, however, that the Armeria plants may produce new flowers throughout the summer and moreover that the plants infested are frequently large, vigorous individuals which are little affected by larval boring. However this may be, it seems that infested plants are generally located in the periphery of larger clusters.

BEMBECIA TRIANNULIFORMIS (Freyer, 1842), comb. n.
Figs. 39, 56, 73, 105.

Sesia triannuliformis Freyer, 1842: 35.

Very similar to B. muscaeformis; for an account of the external characteristics see the treatment of the latter species. FW 6-9 mm. - Male genitalia (Figs. 39, 56): distal end of crista sacculi distinctly thickened, particularly the lower corner. Crista gnathi extending distinctly farther anterior than lateral flaps. Saccus broadest at apex (in dorsal/ventral view).- Female genitalia (Fig. 73): differs from B. triannuliformis in having sclerotized part of ductus bursae straight.

Not recorded from Scandinavia, but the C. European distribution extends to the Baltic, and the species should be sought for in the southern parts of the region. - C. and S. E. Europe to S. W. Asia.

The adults occur in June; they may be found resting on stems foliage or flowers of the food plant, Rumex acetosella. The eggs are laid near the root. The young larva makes tunnels beneath the bark of the primary root. After hibernation it ascends the root stock in a peripheral helical tunnel. In May it makes a tube, 3-5 cm long, stretching from the root stock to the ground surface; the upper end of the tube is closed by a woven lid. Pupation takes place inside the tube. Preferred habitats are sandy fields, dunes, wood glades or road sides.

In May infested plants may be cut below the root stock; they should be planted in a cage and sprinkled. Adult moths may be swept from the food plant.

Note: Rostrup (1896) mentions a Sesiid gall on Rumex acetosella (as Sesia braconiformis H. S. (?)). The gall in question is, however, according to Henriksen (1944) that of a Curculionid beetle, Apion rubiginosum Grill (sanguineum auct. nec. Deg., cf. Hansen, 1965).

Appendix

A brief account is here given of three Sesiid species, which are unrecorded from N.W.Europe, but which might possibly reach the eastern periphery of Fennoscandia.

CHAMAESPHECIA TENTHREDINIFORMIS (Denis & Schiffermüller, 1775)

Sphinx tenthrediniformis Denis & Schiffermüller, 1775: 44.
Sphinx empiformis Esper, 1783: 215.

Type-species of Chamaesphecia s.str. (see p.22, 37). Somewhat similar to Bembecia muscaeformis and B.triannuliformis, but laterofacial scales dark and abdominal bands more prominent, yellow. - C.European distribution extending to Poland and Latvia. - Adult flight period June to July. Larva typically in root stocks of Euphorbia species, particularly E.cyparissus.

CHAMAESPHECIA LEUCOPSIFORMIS (Esper, 1798)

Sphinx leucopsiformis Esper, 1798: 25.

Somewhat similar to Bembecia muscaeformis and B.triannuliformis, but usually larger, laterofacial scales dark, abdomen with transverse band on segment IV only and with prominent longitudinal row of elongated light spots mid-dorsally. - C.European distribution extending to Poland and Latvia. - Adult flight period late: August to September. Larva in root stocks of Euphorbia species, particularly E.cyparissus.

AEGERIA STOMOXIFORMIS (Hübner, 1790)

Sphinx stomoxiformis Hübner, 1790: 93.

A red-banded species, distinguishable by dark laterofacial scales and by having the red band posterolaterally extended to segment VI; inner tegular margin of male broadly red. - C.European distribution extending to Poland and Lithuania. - Adult flight period June to August. Larva in trunks of Mespilus germanica.

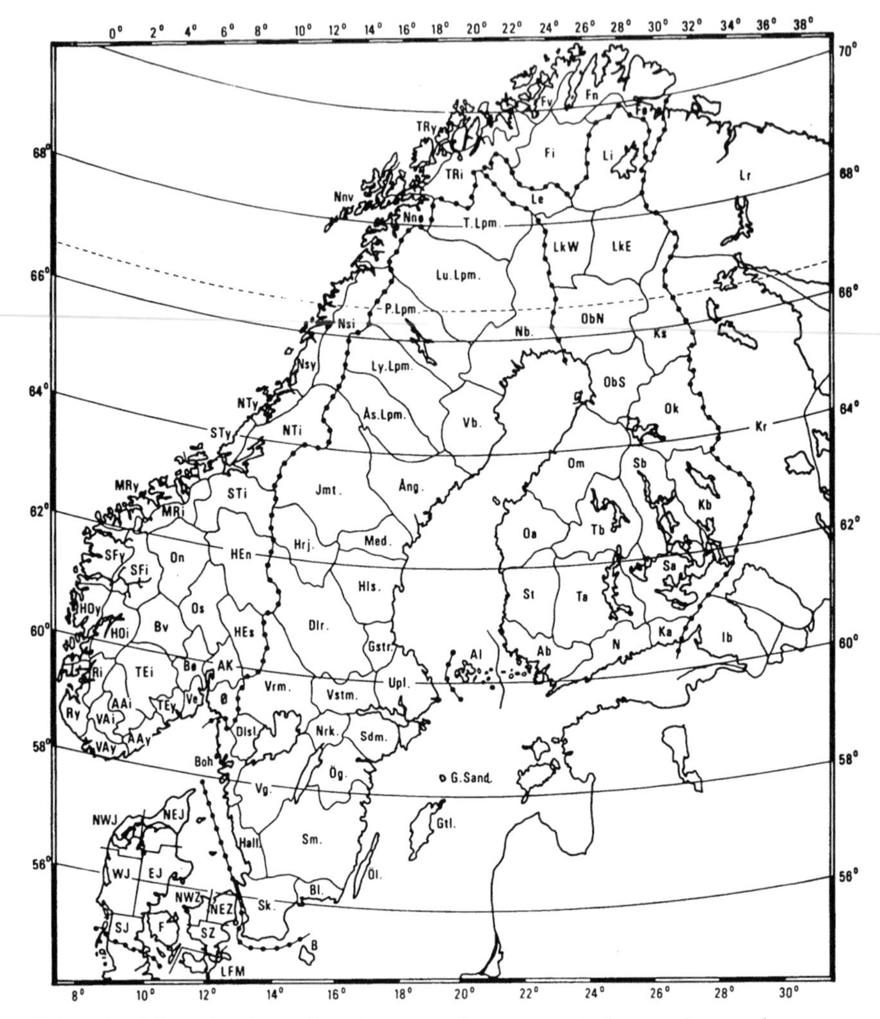

List of abbreviations for the provinces used throughout the text, on the map and in the following tables.

DENMARK

SJ	South Jutland	LFM	Lolland, Falster, Møn
EJ	East Jutland	SZ	South Zealand
WJ	West Jutland	NWZ	North West Zealand
NWJ	North West Jutland	NEZ	North East Zealand
NEJ	North East Jutland	B	Bornholm
F	Funen		

SWEDEN

Sk.	Skåne	Vrm.	Värmland
Bl.	Blekinge	Dlr.	Dalarna
Hall.	Halland	Gstr.	Gästrikland
Sm.	Småland	Hls.	Hälsingland
Öl.	Öland	Med.	Medelpad
Gtl.	Gotland	Hrj.	Härjedalen
G. Sand.	Gotska Sandön	Jmt.	Jämtland
Ög.	Östergötland	Ång.	Ångermanland
Vg.	Västergötland	Vb.	Västerbotten
Boh.	Bohuslän	Nb.	Norrbotten
Dlsl.	Dalsland	Ås. Lpm.	Åsele Lappmark
Nrk.	Närke	Ly. Lpm.	Lycksele Lappmark
Sdm.	Södermanland	P. Lpm.	Pite Lappmark
Upl.	Uppland	Lu. Lpm.	Lule Lappmark
Vstm.	Västmanland	T. Lpm.	Torne Lappmark

NORWAY

Ø	Østfold	HO	Hordaland
AK	Akershus	SF	Sogn og Fjordane
HE	Hedmark	MR	Møre og Romsdal
O	Opland	ST	Sør-Trøndelag
B	Buskerud	NT	Nord-Trøndelag
VE	Vestfold	Ns	southern Nordland
TE	Telemark	Nn	northern Nordland
AA	Aust-Agder	TR	Troms
VA	Vest-Agder	F	Finnmark
R	Rogaland		

n northern   s southern   ø eastern   v western   y outer   i inner

FINLAND

Al	Alandia	Kb	Karelia borealis
Ab	Regio aboensis	Om	Ostrobottnia media
N	Nylandia	Ok	Ostrobottnia kajanensis
Ka	Karelia australis	ObS	Ostrobottnia borealis, S part
St	Satakunta	ObN	Ostrobottnia borealis, N part
Ta	Tavastia australis	Ks	Kuusamo
Sa	Savonia australis	LkW	Lapponia kemensis, W part
Oa	Ostrobottnia australis	LkE	Lapponia kemensis, E part
Tb	Tavastia borealis	Li	Lapponia inarensis
Sb	Savonia borealis		

USSR

Ib Ingria borealis   Kr Karelia rossica   Lr Lapponia rossica

Family SESIIDAE		N. Germany	G. Britain	SJ	EJ	WJ	NWJ	NEJ	F	LFM	SZ	NWZ	NEZ	B	Sk.	Bl.
				DENMARK												
Pennisetia hylaeiformis (Lasp.)	1	●	●	●	●	●	●	●	●	●	●	●	●	●	●	●
Sesia melanocephala Dalm.	2	●			●			●					●		●	
S. apiformis (Cl.)	3	●	●	●	●			●	●	●	●	●	●		●	●
S. bembeciformis (Hübn.)	4	●	●	●	●				●	●			●	●	●	
Paranthrene tabaniformis (Rott.)	5	●	●	●	●			●						●		
Aegeria scoliaeformis (Bkh.)	6	●	●		●			●			●		●		●	●
A. mesiaeformis (H. S.)	7															
A. spheciformis (Den. & Schiff.)	8	●	●	●	●	●		●	●	●	●		●	●	●	
A. andrenaeformis (Lasp.)			●													
A. culiciformis (L.)	9	●	●	●	●		●	●	●	●	●	●	●	●	●	●
A. myopaeformis (Bkh.)	10	●	●												●	
A. formicaeformis (Esp.)	11	●	●	●	●	●	●	●		●	●	●	●	●	●	●
A. polaris (Stgr.)	12															
A. conopiformis (Esp.)																
A. tipuliformis (Cl.)	13	●	●	●	●		●	●	●	●	●	●	●	●	●	
A. spuleri (Fuchs)		●														
A. flaviventris (Stgr.)	14	●	●											●		
A. vespiformis (L.)	15	●	●								●				●	
Bembecia scopigera (Scop.)	16	●	●		●	●	●		●	●	●	●	●	●	●	●
B. chrysidiformis (Esp.)			●													
B. muscaeformis (Esp.)	17	●	●		●	●		●	●	●	●	●	●		●	
B. triannuliformis (Frr.)		●														

SWEDEN

	Hall.	Sm.	Öl.	Gtl.	G. Sand.	Ög.	Vg.	Boh.	Dlsl.	Nrk.	Sdm.	Upl.	Vstm.	Vrm.	Dlr.	Gstr.	Hls.	Med.	Hrj.	Jmt.	Ång.	Vb.	Nb.	Ås. Lpm.	Ly. Lpm.	P. Lpm.	Lu. Lpm.	T. Lpm.
1	●	●	●	●		●	●	●	●	●	●	●	●	●	●	●	●	●	●	●	●	●	●					
2		●	●			●					●	●	●								●							
3	●	●	●	●		●	●	●	●	●	●	●	●	●	●	●	●				●							
4																												
5			●			●	●			●	●	●					●											
6	●	●	●	●		●	●				●	●	●	●	●		●			●		●	●		●		●	●
7																												
8		●	●	●		●	●		●	●	●	●	●	●	●	●	●	●		●	●	●	●		●		●	●
9		●	●	●			●			●	●	●	●	●	●	●	●			●		●	●	●	●	●	●	●
10												●																
11		●	●			●			●	●	●	●		●		●	●	●			●		●	●	●	●	●	●
12																			●				●			●	●	●
13		●	●	●		●	●	●	●	●	●	●	●	●	●	●	●	●		●	●	●	●		●		●	●
14																												
15			●									●																
16	●	●	●	●		●	●	●	●	●	●	●																
17																												

Family SESIIDAE		NORWAY														
		Ø + AK	HE (s + n)	O (s + n)	B (ø + v)	VE	TE (y + i)	AA (y + i)	VA (y + i)	R (y + i)	HO (y + i)	SF (y + i)	MR (y + i)	ST (y + i)	NT (y + i)	Ns (y + i)
Pennisetia hylaeiformis (Lasp.)	1	●	◖	◖			●	●	◖	◖	◗	●			◗	◗
Sesia melanocephala Dalm.	2	◗														
S. apiformis (Cl.)	3	●	●	●	◖	●	◖	◖	◖		●					
S. bembeciformis (Hübn.)	4													◖		
Paranthrene tabaniformis (Rott.)	5	◗					◖									
Aegeria scoliaeformis (Bkh.)	6	●		◗			◗				◗		◗			◗
A. mesiaeformis (H. S.)	7															
A. spheciformis (Den. & Schiff.)	8	●	◖	●	◖	●					◗	◖	◖		◗	◖
A. andrenaeformis (Lasp.)																
A. culiciformis (L.)	9	◗	◖	◖	◖	●	◗				●			◗		◗
A. myopaeformis (Bkh.)	10	●								◖						
A. formicaeformis (Esp.)	11	◗				●				◖						
A. polaris (Stgr.)	12			●	◗											
A. conopiformis (Esp.)																
A. tipuliformis (Cl.)	13	●		◖	◖	●		◖		◖	●	◗				
A. spuleri (Fuchs)																
A. flaviventris (Stgr.)	14															
A. vespiformis (L.)	15															
Bembecia scopigera (Scop.)	16	◗			◖											
B. chrysidiformis (Esp.)																
B. muscaeformis (Esp.)	17															
B. triannuliformis (Frr.)																

	Nn (ø + v)	TR (y + i)	F (v + i)	F (n + ø)	FINLAND																				USSR		
					Al	Ab	N	Ka	St	Ta	Sa	Oa	Tb	Sb	Kb	Om	Ok	Ob S	Ob N	Ks	LkW	LkE	Le	Li	Ib	Kr	Lr
1					●	●	●	●	●	●	●	●	●	●	●	●		●							●	●	
2					●	●	●	●	●			●			●	●										●	
3					●	●	●	●	●	●	●		●	●	●			●	●						●	●	
4														●							●	●					
5					●	●	●	●	●	●	●				●										●	●	
6		●			●	●	●	●	●	●	●	●	●	●	●	●	●				●	●		●	●	●	
7						●	●	●																			
8	◖				●	●	●	●	●	●	●	●	●	●	●	●	●	●	●						●	●	
9	◖	●	◗	◗	●	●	●	●	●	●	●	●	●	●	●	●	●	●	●	●	●	●	●	●	●	●	
10																											
11					●	●	●	●	●	●	●	●		●	●	●		●	●		●	●			●	●	
12																				●			●	●			
13					●	●	●	●	●	●	●	●	●	●	●	●		●	●						●	●	
14						●	●		●	●					●												
15																											
16					●	●	●																				
17																											

Literature

Bartel, M., 1912: Aegeriidae (Sesiidae), in: Seitz, A., Die Grossschmetterlinge der Erde 2: 365-416, pl. 51, 52. Stuttgart.

Bergmann, A., 1953: Die Grossschmetterlinge Mitteldeutschlands 3. Spinner und Schwärmer. XII + 551 pp., 56 pls. Jena.

Bradley, J.D., Fletcher, D.S. & Whalley, P.E.S., 1972: Lepidoptera. In Kloet, G.S. & Hincks, W.D., A Check List of British Insects (Ed. 2) - Handbk Ident. Br. Insects 11 (3). 153 pp.

Beutenmüller, W., 1901: Monograph of the Sesiidae of America, North of Mexico. - Mem. Am. Mus. nat. Hist., 1, 6: 215-352, pl. 29-36.

Borkhausen, M.B., 1789: Naturgeschichte der Europäischen Schmetterlinge nach systematischer Ordnung. 2. Schwärmer. 4+96+239 pp., 1 pl. Frankfurt.

Brock, J.P., 1971: A contribution towards an understanding of the morphology and phylogeny of the Ditrysian Lepidoptera. - Jnl. nat. Hist., 5: 29-102.

Börner, C., 1939: Die Grundlagen meines Lepidopterensystems. - Verh. VII Int. Kongr. Ent. 1938, 2: 1372-1424.

- 1959: Lepidoptera, Schmetterlinge, in P. Brohmer, Fauna von Deutschland. (Ed. 8): 382-421.

Capuse, I., 1973: Zur Systematik und Morphologie der Typen der Sesiidae (Lepidoptera) in der R. Püngeler-Sammlung des Zoologischen Museums zu Berlin. - Mitt. münch. ent. Ges., 63: 134-171.

Classey, E.W., Cockayne, E.A., Fassnidge, W., Fletcher, J.B., Parfitt, R.W. & Tams, W.H.J., 1946: Collecting Clearwings. - Leaflet 18. Amateur Entomologists' Society. 12 pp.

Clerck, C.A., 1759: Icones insectorum rariorum cum nominibus eorum trivialibus locisque e. C. Linnaei syst. nat. allegatis. 1. 8+16 pls. Stockholm.

Common, I.F.B., 1969: Lepidoptera, in CSIRO (sponsor), The Insects of Australia: 765-866. Melbourne.

Dalman, I.W., 1816 (-1817): Försök til systematisk uppställning af Sveriges Fjärilar. - K. Vet. Ac. Handl., 37: 48-101, 199-225, 2 pls.

Denis, M. & Schiffermüller, I., 1775: Ankündung eines systematisches Werkes von den Schmetterlingen der Wienergegend. 323 pp., 2 pls. Vienna.

Edelsten, H.M., Fletcher, D.S. & Collins, R.J., 1961: Ed. 3 of R. South: The Moths of the British Isles. 2. 379 pp., 141 pls. London.

Engelhardt, G.P., 1946: The North American clear-wing moths of the family Aegeriidae. - Bull. U.S. natn. Mus. 190, VI + 222 pp., 32 pls.

Esper, E.J.C., 1779-86, 1789-1806: Die Schmetterlinge in Abbildungen nach der Natur mit Beschreibungen 2. 234 pp., 36 pls. Supplementum 52 pp., 4 pls. (Dates according to Sherborn, C.D. & Woodward B.B. Ann. Mag. nat. Hist., VII, 7: 137-140, 1901).

Forster, W. & Wohlfahrt, T.A., 1960: Die Schmetterlinge Mitteleuropas 3 (Spinner und Schwärmer). VIII + 239 pp., 28 pls. Stuttgart.

Freyer, C.F., 1842: Neuere Beiträge zur Schmetterlingskunde mit Abbildungen nach der Natur 5. 166 pp., 96 pls. Augsburg.

Fuchs, J., 1908: Sesia spuleri nov. spec. - Int. ent. Z. 2: 33.

Gullander, B., 1963: Nordens svärmare och spinnare. VII + 104 pp. Stockholm.

Hansen, V. 1965: Biller XXI, Snudebiller. Danmarks Fauna 69. 524 pp. Copenhagen.

Henriksen, K. L., 1944: Fortegnelse over de danske Galler (Zoocecidier) - Spolia Zool. Mus. Hauniensis 6, 212 pp.

Hering, M., 1932: Die Schmetterlinge nach ihren Arten dargestellt. - Tierwelt Mitteleur., Ergänzungsband 1. 545 pp. Leipzig.

Herrich-Schäffer, G. A. W., 1845: Systematische Bearbeitung der Schmetterlinge von Europa, als Text, Revision und Supplement zu J. Huebners Sammlung europäischer Schmetterlinge 2. Schwärmer, Spinner, Eulen. 140 pp., 190 pls. Regensburg.

Hoffmeyer, S., 1960: De danske spindere. Ed. 2. 270 pp., 24 pls. Århus.

Hübner, J., 1790: Beiträge zur Geschichte der Schmetterlinge. 2. 128 pp., 16 pls. Augsburg.

- 1797: Sammlung europäischer Schmetterlinge 2. Sphinges, Schwärmer. 32 + 4 pp., 38 pls. Augsburg.

Kaaber, S., 1964: Om nødvendigheden af kildekritik i entomologisk faunistik - Flora og fauna 70: 45-56.

Kemner, N. A., 1922: Zur Kenntnis der Entwicklungsstadien einiger Sesiiden. - Ent. Tidskr. 1922: 41-57.

Kristensen, N. P., in press, a: Sesia andrenaeformis Laspeyres, 1801, the specific name to be protected against the nomen oblitum Sesia anthraciniformis Esper, 1798 (Lepidoptera: Sesiidae). - Bull zool. Nom.

- in press, b: Sphinx tipuliformis Clerck, 1759, the specific name to be protected against the nomen oblitum Sphinx salmachus Linnaeus, 1758 (Lepidoptera: Sesiidae). - Ibid.

Lampa, S., 1883: Anteckningar om sällsyntare svenska Lepidoptera. - Ent. Tidskr., 1883: 125-128.

Laspeyres, J. H., 1801: Sesiae Europaeae iconibus et descriptionibus illustratae. 32 pp., 1 pl. Berlin.

Lewin, J. W., 1797: Observations respecting some rare British Insects. - Trans. Linn. Soc. Lond., 3: 1-4, 2 pls.

Linné, C., 1758: Systema naturae. Per regna tria naturae. Ed. 10., 1, 824 pp. Stockholm 1758.

- 1761: Fauna svecica sistens animalia Sveciae regni. Ed. 2, 578 pp. Stockholm.

Mc Kay, M., 1968: The North American Aegeriidae (Lepidoptera): A revision based on late-instar larvae. - Mem. entomol. Soc. Can., 58, 112 pp., 49 + figs.

Naumann, C., 1971: Untersuchungen zur Systematik und Phylogenese der holarktischen Sesiiden (Insecta, Lepidoptera). - Bonner zool. Monogr., 1: 190 pp.

Niculescu, E. V., 1962: Contributions à l'étude morphologique des Aegeriides (Lepidoptera) paléarctiques. III- Aegeria ignicolle Hmps. - Bull. Soc. ent. Mulhouse, 1962: 33-37.

Niculescu, E. V., 1964: Les Aegeriidae. Systematique et phylogenie. - Linn. Belg., 3: 34-45.

Nordman, A. F., 1963: Zur Biologie und Verbreitung von Aegeria melanocephala Dalm. (Lep., Aegeridae). - Mem. Soc. Fauna Flora fenn. 39: 143-158.

Nordström, F., Opheim, M. & Sotavalta, O., 1961: De Fennoskandiska Svärmarnas och Spinnarnas Utbredning. - Lund. Univ. Årsskr. N. F. Afd. 2, 57 (4), 87 pp., 181 maps. Lund.

Nordström, F. & Wahlgren, E., 1941: Svenska Fjärilar. 353 pp., 50 pls. Stockholm.

Popescu-Gorj, A. & Capuse, I., 1969: Revision de Paranthrene tabaniformis Rott. (Lepidoptera Aegeriidae) et des especes europeennes apparentees. - Bull. mens. Soc. linn. Lyon, 9: 315-328.

- Niculescu, E. & Alexinschi, A., 1958: Lepidoptera Familia Aegeriidae. - Fauna Repub. pop. rom., Insecta 11, 1. 195 pp., 5 pls. Bukarest.

Roovers, M., 1964: The genitalia of the Dutch Aegeriidae. - Zool. Meded., 40: 97-113.

Rostrup, S., 1896: Danske Zoocecidier. - Vidensk. Meddr dansk naturh. Foren., 48: 1-64.

Rothschild, N. C., 1906: Notes on the Life History of Trochilium andrenæforme, Lasp. - Trans. ent. Soc. Lond., 1906: 471-473.

Rottemburg, S. A. von., 1775: Anmerkungen zu den Hufnagelischen Tabellen der Schmetterlinge 2. - Der Naturforscher, 7: 105-112.

Saramo, Y., 1973: Synanthedon mesiaeformis H. S. (Lep., Aegeriidae) at Kotka and in the eastern archipelago of the Gulf of Finland. - Annls. Ent. Fenn., 39: 24-28.

Scopoli, J. A., 1763: Entomologia carniolica exhibens insecta carnioliae indigene et distributa in ordines, genera, species, varietates methodo Linnaeana. 421 pp.

Schantz, M. v., 1959: Studien über Synanthedon polaris Stgr. - Not. Ent. 39: 33-43.

Schwarz, R. & Niculescu, E., 1962: Morphologische, biologische und systematische Beiträge zur Kenntnis der Aegeriidae (Lepidoptera). - Z. Arbeitsgem. östr. Ent., 14: 42-47.

Spuler, A., 1910: Die Schmetterlinge Europas 2, 523 pp. Stuttgart.

Staudinger, O., 1877: Neue Lepidopteren des europäischen Faunengebietes aus meiner Sammlung. - Stettin. ent. Ztg., 38: 175-208.

- 1883: Einige neue Lepidopteren Europa's. - Stettin. ent. Ztg., 44: 177-186.

Tuxen, S. L. (ed.), 1970: Taxonomist's Glossary of Genitalia in Insects. Ed. 2. 369 pp. Copenhagen.

Urbahn, E. & H., 1939: Die Schmetterlinge Pommerns mit einem vergleichenden Überblick über den Ostseeraum. - Stettin. ent. Ztg., 100: 185-826.

Valle, K. J., 1937: Suurperhoset. Macrolepidoptera 2. Kiitäjät, Sphinges ja Kehrääjät, Bombyces. - Suomen Eläimet, Animalia Fennica 3. 213 pp., 13 pls. Helsinki.

Authors' adresses:

M. F.: Skyggelundsvej 8,
2500 Valby, Denmark

N. P. K.: Zoological Museum,
Universitetsparken 15,
Copenhagen 2100 Ø, Denmark

Index

Page references to the key and to the main taxonomic treatment. Valid names are underlined.

Printed in the United States
By Bookmasters